GREEN YOGA

Georg Feuerstein
and Brenda Feuerstein

Traditional Yoga Studies
Saskatchewan

Published by
Traditional Yoga Studies
Box 661, Eastend, Saskatchewan S0N 0T0, Canada
www.traditionalyogastudies.com — tyslearning@sasktel.net

First Edition. Printed by Houghton Boston in Saskatoon, Canada. This edition is printed on ancient forest friendly (AFF) 100% post-consumer recycled paper and with vegetable-based, low-VOC inks.

Library and Archives Canada Cataloguing in Publication

Feuerstein, Georg
Green yoga / Georg Feuerstein and Brenda Feuerstein.

Includes bibliographical references and index.
ISBN 978-0-9782138-2-4

1. Yoga. 2. Ecology--Religious aspects--Hinduism. 3. Nature--Religious aspects--Hinduism. I. Feuerstein, Brenda, 1963- II. Title.

B132.Y6F465 2007 294.5'436 C2007-901937-4

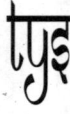

Traditional Yoga Studies is a Canada-registered company specializing in promoting traditional wisdom, particularly Yoga in its various forms and branches, through distance-learning programs, publications, and a web site. **Green Yoga Initiative** is a program offered through TYS, which is designed to educate the public, especially Yoga practitioners, about the present environmental crisis and how to overcome it. A *free* subscription to GYI's e-newsletter is available. GYI encourages the formation of local action groups based on the book *Green Yoga.*

DEDICATION

We gratefully dedicate this book

. . . to Arne Naess (1912–),
originator of Ecosophy and philosopher extraordinaire,
who for over six decades has been an unceasing source
of inspiration and a resounding voice of sanity
for worldwide peace, sustainability, and joy

AND

. . . to David Takayoshi Suzuki (1936–),
who since the 1960s has been a most valiant spokesman
for the environment and human responsibility
and whose many publications and educational documentaries
have both broadened and deepened our understanding
of the unique and perilous time in which we live.

I am confident that humans have what is demanded to turn things around and achieve Green societies.*

—Arne Naess

We're in a giant car heading towards a brick wall and everyone's arguing where they're going to sit.**

—David Suzuki

Modern society will find no solution to the ecological problem unless it *takes a serious look at its life style.* In many parts of the world society is given to instant gratification and consumerism while remaining indifferent to the damage which these cause. As I have already stated, the seriousness of the ecological issue lays bare the depth of man's moral crisis.***

—H. H. Pope John Paul II

Ultimately, humanity is one and this small planet is our only home. If we are truly to help one another and protect this home of ours, each of us needs to experience a vivid sense of compassion and responsibility.†

—H. H. The Dalai Lama

---

*"Deep Ecology For the Twenty-Second Century," in *Deep Ecology For the 21st Century,* ed. by George Sessions (Boston, Mass.: Shambhala Publications, 1995), p. 467.

**David Suzuki,as cited by *Brainy Quotes.*

***Message of His Holiness Pope John Paul II For the Celebration of the World Day of Peace, 1 January 1990, Section IV.13.

†The Dalai Lama, "The True Source of Political Success," in Stephanie Kaza and Kenneth Kraft, eds., *Dharma Rain: Sources of Buddhist Environmentalism* (Boston, Mass.: Shambhala Publications, 2000), p. 167.

CONTENTS

The change that is taking place on the earth and in our minds is one of the greatest changes ever to take place in human affairs, perhaps *the* greatest, since what we are talking about is not simply another historical change or cultural modification, but a change of geological and biological as well as psychological order of magnitude.*

—Thomas Berry

We are standing on the cusp of a very precarious future, as immediate as the lifetime of our children. There are those who want to swap it for their short-term gain, and they will if we let them. But we know too much to lay the blame on them. If we passively yield control of our collective destiny to those destroyers, we will all share the responsibility for the unspeakable consequences. If each of us chooses inaction, it will make us all accomplices in the triumph of greed and short-sightedness and selfishness.**

—Ross Gelbspan

One of the questions I am frequently asked when I am speaking in various countries is, Given the environmental problems that the world is facing, can we make it" . . . My answer is always the same: it depends on you and me, on what you and I do to reverse these trends. It means becoming politically active. Saving our civilization is not a spectator sport.***

—Lester R. Brown

---

*Thomas Berry, *The Dream of the Earth* (San Francisco: Sierra Club Books, 1988), pp. 11–12.

**Ross Gelbspan, *The Heat Is On: The High Stakes Battle Over Earth's Threatened Climate* (Reading, Mass.: Addison-Wesley, 1997), p. 195.

***Lester R. Brown, *Plan B 2.0* (New York: W. W. Norton, 2006), p. 264.

PREFACE

Sixty-five million years ago, a large asteroid crashed into the Yucatan peninsula and triggered worldwide repercussions that led to the extinction of an estimated 70 percent of all existing species at the end of the Cretaceous Period, including the giant dinosaurs.* Biologists speak of that tragedy as the Fifth Mass Extinction, suggesting that there have been other catastrophes of this magnitude in Earth's past.

Today, the Sixth Mass Extinction is happening under our very noses! Every day, an estimated 150 species are becoming extinct. We are talking here about entire *species* comprising tens of thousands and even millions of individual plants, insects, and animals.

Astonishingly, most people are ignorant of the fact that they are living in fateful days. To put it bluntly: our planet is dying, and we are turning a blind eye to it!

A 1998 survey of 5,000 scientists revealed that 3,500 believed that we are witnessing a biological catastrophe of unprecedented proportions. What is happening today far outstrips the calamity that spelled the sudden demise of the dinosaurs.

This time round, however, the culprit is definitely not an extraterrestrial rock of some seven miles in diameter smashing

*For a well-documented and accessible account of past mass extinctions, see Prof. Trevor Palmer's *Perilous Planet Earth: Catastrophes and Catastrophism Through the Ages* (Cambridge: Cambridge University Press, 2003).

into the Earth. Rather, the cause can be found on our home planet itself. More precisely, it is our own populous human species that is having such a devastating impact on the biotic environment. Carbon pollution, rising temperatures of air and water, deforestation and desertification, as well as pollution of the ocean, lakes, rivers, groundwater, and the air combine to destabilize natural systems. Nature has begun to recalibrate itself: the polar ice caps and glaciers are melting rapidly, ocean levels are rising and are threatening entire island nations and coastal cities, and megahurricanes are causing numerous deaths and widespread multibillion-dollar destruction.

Prof. Paul Ehrlich at Stanford University, who first warned us about the dire consequences of the population explosion, recently reiterated the warning he had sounded decades earlier: We must anticipate billions of people dying from hunger and thirst.*

There is no telling where the present mass extinction will stop, but increasingly biologists worry that humanity may well end up among its casualties before it's all over. This should make the headlines every day, but it is not. Governments should impose stringent measures on industry, business, and society as a whole, but they are not. Educators should make a priority of this gruesome reality, but they are not. Ignorance and lethargy prevent concerted action. It seems that humanity is sleepwalking into oblivion.

The question is what can and must we do? Specifically, how ought a spiritually engaged person respond to this grave situation? Even more specifically, what ought a Yoga practitioner do?

In this book, we will not oh so gently nudge you in the

---

*See Paul Ehrlich's short statement in the online video clip "Call of Life: Facing the Mass Extinction," featured on the website of Species Alliance at www.speciesalliance.org.

right direction. We believe that it is way past the time for mild reminders. This is intended as a last-minute wake-up call.

* * * * *

Yoga is practiced today by many millions of Westerners. Most of them seek to benefit exclusively from Yoga's physical exercises, which demonstrably have great efficacy as therapeutic and preventative measures. As we will argue in this book, however, the moral and spiritual teachings of this 5,000-year-old tradition are even more potent and significant.* They can, we believe, not only serve as powerful tools of self-transformation but also can help us deal with our present-day worldwide crisis more appropriately, responsibly, and effectively.

That we live at a time of decisive crisis is ever so slowly dawning on those whose intelligence has not succumbed to slumber. Our crisis, to be sure, is not only of an environmental nature. It affects virtually all aspects of human existence, including our social, political, moral, and spiritual life.

In the following, we will touch on all these aspects but of necessity focus only on their most salient features. Foremost in our deliberations will be the environment, but we cannot talk about the environment without also talking about lifestyle, consumerism, economics, politics, cultural blindspots, ethics, values, and worldview.

Arne Naess, the Norwegian philosopher and originator of Ecosophy, independently voiced the same opinion. In an interview conducted in 1982 for *The Ten Directions* magazine, he

---

*See Georg Feuerstein, *Yoga Morality: Ancient Teachings At a Time of Global Crisis* (Prescott, Az.: Hohm Press, 2007). In many ways, *Green Yoga* is a companion volume to *Yoga Morality*.

prophetically observed:

> Within fifty years, either we will need a dictatorship to save what is left of the diversity of life forms, or we will have a shift of values, a shift of our total view such that no dictatorship will be needed.*

To the above, we would like to add the strong words of the internationally renowned Canadian educator David Suzuki, host of the popular television program "The Nature of Things," which has been running since 1960:

> My primary concern continues to be to convince people that there is a real crisis affecting all of us and that every minute we continue to deny or ignore it, the fewer options we will have and the harder it will be to make the changes needed to bring us back into balance with the factors that sustain our lives.**

Now, traditional Yoga commends itself for at least two reasons. First, it champions a holistic worldview, which combines ethics, psychology, and philosophy with what for lack of a better word we would like to call "mind training."

Second, Yoga is inherently "green." That is to say, the core values and practices of traditional Yoga coincide with the values

---

*Arne Naess cited by Stephan Bodian, "Simple in Means, Rich in Ends," reprinted in George Sessions, ed., *Deep Ecology For the 21st Century* (Boston and London: Shambhala Publications, 1995), p.26. We are grateful to Arne Naess, who is now in his 90s, for graciously accepting our tribute to him in the form of our dedication on page 4 of this book.

**David Suzuki, *Time To Change: Essays* (Toronto: Stoddart Publishing, 1994), pp. xi-xii.

promoted by the Deep Ecology movement. Both advocate reverence for all life forms.

In case you were wondering, Green Yoga in the Sanskrit language in which most Yoga texts are written is *harid-yoga,* which almost sounds like "heart Yoga." *Green* does indeed mean having a heart and being whole-hearted about acknowledging the fact that we are blood relations to all life wherever it may appear on our planet or in the universe as a whole. Although our DNA may not owe anything to extraterrestrial life forms, we all are made of the same basic stuff, which some have poetically called "star dust."

In this work, then, we will argue that the moral and philosophical teachings of Yoga, as well as its spiritual practices, are as relevant today as they were thousands of years ago. If anything, they are even more pertinent today than they were in the infancy of our human civilization. In particular, they have much to tell us about a hale relationship to Nature.

We endorse the American novelist William Burroughs's astute observation found in his book *The Western Lands* that "desperation is the raw material of drastic change." We simply must recognize that our situation is in fact desperate. In any case, change we will, either voluntarily or under the compulsion of events!

After a lifetime of warning people of the great seriousness of the situation, David Suzuki in a recent interview remarked:

> I don't know what it will take to get people to realize this is not a joke; this is not speculation. We should be in crisis mode.*

---

*See David Suzuki quoted in Swami Sivananda, "This Man Can't Save You," *Ascent Magazine,* no. 29 (Spring 2006).

We agree entirely if "crisis mode" means that we should be alert and wholeheartedly committed to well-considered remedial action.

The extraordinary challenges we are facing today can all be attributed to humanity's long-standing failure to live wisely and responsibly. But what is humanity if not each individual? Thus, our worldwide crisis is a vivid reminder that we all must become responsible for our actions, as well as for our refusal to act where action is called for in order to prevent damage or harm.

As we see it, the problems in the world today are forcing us to grow up, both individually and as a species. If we fail to achieve emotional, intellectual, moral, and spiritual maturity, we will inevitably contribute to the wreckage of our planet and the possible (and even likely) demise of the human species. This somber conclusion is shared by a growing number of people, especially those who work in the fields of environmental studies, biology, sociology, and politics.

Although *Green Yoga* is written especially for practitioners of Yoga, its message is pertinent to everyone alive today. **Yoga practitioners, we feel, have a particular obligation to live in a green way, mainly because this reflects the essence of the tradition which they espouse.** We realize that few adherents of contemporary Yoga have, as yet, discovered Yoga's spiritual and moral teachings beyond the allure of the bodily postures. But it would be too uncharitable to think that they could not do so in the future. Yoga has a power all its own, and not a few people have gradually moved from mere posture practice to a deep commitment to yogic morality and spirituality.

Furthermore, whatever level one's involvement with Yoga may be, we all are facing the same environmental and social problems and sooner or later will want to find our own personal response to them. For those who are still in the process of discovering Yoga's spiritual and moral depth, this book can serve as a signpost. Besides, everyone is able to do something *now*.

Contemporary Yoga practitioners are clearly demonstrating their capacity to act responsibly. They are acting responsibly toward their own body by taking good care of it. The mind, of course, does not become automatically sound when the body's health is taken care of, though physical well-being helps.

In addition, however, we must strive toward creating a healthy mind for ourselves. In Yoga, this is achieved by harmonizing our actions through proper morality and by quieting the mind through regular meditation practice.

Some yogic approaches favor working exclusively with the mind on the understanding that when the mind becomes clear and "sound," we will be able to have the right relationship to our embodiment. However, Hatha-Yoga espouses a distinctly different orientation, which is based on the assumption that we can achieve optimal physical health and even create the kind of body that is a perfect instrument for spiritual realization. We will not argue the merit of these two positions here. What is important is that both aim at liberating us from the limitations of our human condition by discovering our true nature as immortal Spirit, or Being-Awareness. One approach simply does this via the mind, and the other via the body.

Contemporary Yoga practitioners are now called to expand their circle of responsibility by including in their daily practice sound attitudes toward the environment, Mother Nature. To be truly valid today, contemporary Yoga must become as green as the meadows of Ireland or the dense, if rapidly vanishing, rain forests of Borneo.

One vital question ought to be addressed up front: In the face of all the problems talked about in this book, how can we avoid succumbing to mere frustration, anger, sadness, or grim determination? If we are sensitive and in full grasp of the sobering realities of our era, it is indeed not easy to remain level headed. Yoga teaches that whatever our situation may be, we always have the potential for inner harmony and joy. In fact, through stable spiritual practice, we can and must find that still place within us from which wise and compassionate action can flow.

If you would like to receive regular updates on the state of affairs along with positive practical advice or are interested in forming a local action group, you might like to check out our Green Yoga Initiative program, which includes a free informative e-newsletter.

May we all find peace and happiness and co-create a benign future for all beings!

Georg and Brenda Feuerstein

ACKNOWLEDGMENT

We would like to acknowledge all those intrepid researchers, scientists, writers, and whistle-blowers who have dared to swim against the stream in order to tell the truth about the state of the world, often at great personal risk and hardship.

We would also like to thank our spiritual mentor, Neten Rinpoche, for wholeheartedly supporting and blessing this project, responding quickly with skillful means to the concerns we raised in light of our work and often harrowing findings, and for never failing to be a compassionate, healing bodhisattva presence for all his students and the world at large.

1

AS WITHIN, SO WITHOUT

Our Planet Is Dying—What Will YOU Do?

In case you skipped our Preface, as not a few readers are wont to do, we repeat our initial statement here: Every day, approximately 150 entire *species* are said to vanish forever. We are in the midst of the Sixth Mass Extinction, as biologists have labeled this immeasurable tragedy. Our planet is dying—not slowly but more rapidly than even pessimistic scientists predicted only a few years ago! Do we need to spell it out? If this trend continues unabated, our own species will inevitably suffer the same fate before very long.

We are talking decades rather than centuries. You can do the math for yourself: There is an estimated total of about ten million species (most of which are insects!). Divide this by 54,750 (=150/day x 365 days), and you get 182 years. Some biologists think that there may be only 3 million species, which would lower the time frame to c. 55 years. This does not take into account the fact that the extinction rate is accelerating and will become a veritable avalanche as the devastation proceeds.

The fact that biologists do not have precise figures for the number of species in existence or of species becoming extinct every day does not mean that they may be wrong about the threat to life on Earth. Scientists generally are a conservative bunch. So, when seven out of ten biologists believe that we are facing a mass extinction today, then we should take this very seriously indeed. In March 2005, a group of more than 1,300 scientists from 95 countries completed the Millennium Ecosystem Assessment project. In their report, we can read:

> [A]pproximately 60% (15 out of 24) of the ecosystem services examined during the Millennium Ecosystem Assessment [2001–2005] are being degraded or used unsustainably, including fresh water, capture fisheries, air and water purification, and the regulation of regional and local climate, natural hazards, and pests. . .
>
> The number of species on the planet is declining. Over the past few hundred years, humans have increased the species extinction rate by as much as 1,000 times over background rates typical over the planet's history. . .*

Considering that the news media thrive on bad news, this report and others like it should have made the headlines worldwide, but it didn't.

Now, let's consider oxygen production, which is no small matter for life on Earth. The phytoplankton in the oceans is said to generate about 70 percent of the oxygen in the air we breathe. The remaining 30 percent of oxygen stem from forests and grasslands.

The problem is that we are mowing down forests as if

*Millennium Ecosystem Assessment, *Ecosystems and Human Well-being: Synthesis* (Washington, DC.: Island Press, 2005). This report can be read in full online at www.maweb.org.

there were no tomorrow, and grasslands too are suffering from overgrazing, pollution, and desertification. Sadly, the phytoplankton floating in the open ocean, which consists of trillions of microganisms, is also dying. So-called (oxygen-starved) "dead zones" in the oceans are on the increase, as we will explain in Chapter 3.

Here comes the clincher. Some biologists estimate that even a reduction of a mere 5 percent in the total oxygen production could destabilize the entire biotic system on Earth.

If true and being realistic about it, we might have just a few decades left before humanity could be added to the list of extinct species, except there would be no one around to record our species' demise. Given the proverbial lethargy of people and governments and in view of the often sinister ways of corporations, which pursue the profit motive come what may, this is not very much time to turn things around. In other words, there is no time to waste.

Governments, corporations, a few academic institutions, and a handful of private citizens have started to worry about the phenomenon of Peak Oil.* This concept was introduced to demarcate the point at which oil production can no longer keep up with oil consumption. Since our modern civilization is a voracious oil consumer and indeed revolves around oil, some think that Peak Oil spells the imminent collapse of civilization as we know it. It is now widely thought that this point will be reached somewhere between 2015 and 2030. So, conceivably, the collapse of industrial civilization and of the environment will more or less coincide.

Given the fast-paced breakdown of our biotic environment, Peak Oil is beginning to look like a rather localized

---

*See Richard Heinberg, *The Party's Over: Oil, War and the Fate of Industrial Societies* (Gabriola Island, British Columbia: New Society Publishers, 2003).

concern that affects the human species and primarily the wealthy nations. The Sixth Mass Extinction is a phenomenon of far greater scope and consequence. It is likely to outpace the aftermath of Peak Oil. That is to say, "Omega Point," when planetwide extinction becomes inevitable, will likely be reached before the last drop of oil has been extracted.

If you care about the continuation of our human species and life on Earth in general, you must NOW do your utmost to transform your own life through a drastic lifestyle change and to vigorously help transform the way governments and corporations go about their business.

Don't wait for a better moment—tomorrow, in a month, or next year. Don't wait for someone else to take the initiative. Don't think that your personal efforts won't matter. All this is wrong thinking, or what the Yoga adepts of bygone ages called *mithya-darshana,* "flawed vision."

If you are still young enough to expect to live another twenty or thirty years, your own life might well be cut short by the increasing devastation of the biosphere. If you are in your middle age, think of your children and grandchildren whose life will be shortened. No, they will not even inherit a wasteland, because they will likely die of oxygen starvation, overexposure to ultraviolet rays from the Sun, hunger, thirst, or flooding first. (Ultraviolet radiation will increase as the oxygen mantle around the Earth becomes thinner.)

This may all sound very dramatic. However, our collective situation *is* dramatic. Back in the 1960s when the first ecologists were warning the public of the baneful effects of environmental pollution, we could perhaps still afford to ignore them. By the 1980s, we should definitely have started to heed the chorus of environmental voices. Today is an altogether different moment.

We derive no joy from having to be messengers of bad news. The facts are overwhelming: We are clearly destroying our planet's biosphere at a rapid and accelerating rate. **We may have just a few years left to take the most drastic measures as a species, not merely as individuals or even as a single nation.** The present crisis is something to be very scared about.

We must not, however, allow ourselves the luxury of being so petrified that we stand idly by, waiting for doomsday. We must act, and we must act *now*! Is there still time to redirect the tide of events? It all depends on what *everyone* will do and what, at this point, Nature has in mind. Our own generation may never see the benefits of such a collective turn-about. In an attempt to balance itself, Nature has already mobilized forces to countermand the continuing stresses on the environment.

Regardless, we owe it to our children and future generations and indeed to all life forms on this Earth, to take incisive action *immediately* in an attempt to prevent the worst from happening. As responsible Yoga practitioners, we really have no other reasonable option.

In this book, we will make plentiful suggestions for your self-transformation and your participation in sociocultural transformation and corrective environmental action.

Do something constructive! Start now! If you are a spiri-tual practitioner and rational, you really have no choice. Even if you don't consider yourself a spiritual practitioner but simply love life or your children and grandchildren, you must still act responsibly and promptly. You must make your thinking and your lifestyle radically green.

We all must now embrace the Earth as dearly as we embrace our own life, and our love must be charged with great wisdom and compassion for all and with the will to make a difference for future generations while we still can.

Yoga For Purification

There is little doubt: Yoga is India's greatest gift to the world.* Its contribution to our understanding of the human condition and the spiritual potential of our species is inestimable.

Yoga comprises many approaches. It is useful to distinguish between three *forms* of traditional Yoga—Hindu, Buddhist, and Jaina Yoga. These can be said to constitute the spiritual core of the three great cultural complexes of Hinduism, Buddhism, and Jainism. Within these three forms, we can differentiate between various *types* or *branches*. Hindu Yoga, which will be our main reference point in this book, encompasses six major branches, namely Raja-Yoga (the path of meditation), Hatha-Yoga (the path of spiritualizing the body), Jnana-Yoga (the path of wisdom), Karma-Yoga (the path of self-transcending action), Bhakti-Yoga (the path of devotion), Mantra-Yoga (the path of sacred sound), and Tantra-Yoga (the path of psychophysical integration).

It is estimated that there are some thirty million Yoga practitioners in North America. If we were to include Yoga practitioners in other parts of the world, this figure would probably have to be doubled. This represents a considerable number of people, which, among other things, should put to rest any notion that Western Yoga, which was launched in the late nineteenth century, is a mere passing fad. Yoga counts!

Most contemporary Yoga practitioners, including many Westernized Indians, are interested exclusively in what has been

---

*For a detailed discussion of the various forms and types of Yoga, see Georg Feuerstein, *The Yoga Tradition* (Prescott, Ariz.: Hohm Press, 2d ed. 2001). This is the single most comprehensive work on the history, philosophy, and literature of Yoga available today.

called Modern Postural Yoga,* which is a largely secularized derivation of traditional Hatha-Yoga. It is primarily concerned with physical fitness, health, and mental relaxation as part of a program of combating the negative health and psychological effects of modern life.

Thus, Westernized Hatha-Yoga has the goal of restoring or maintaining one's bodily well-being. It seeks to accomplish this chiefly through extensive postural practice and to a lesser extent also through breath control and meditation. The spiritual goals of traditional Hatha-Yoga, such as the cultivation of higher states of consciousness and ultimate liberation, regrettably play only a minor role in Modern Postural Yoga.

Pure Like a Lotus in a Muddy Pond

There are many ways of looking at Yoga. We can regard it as the single most comprehensive consciousness technology ever devised. Or we can understand it as the oldest tradition of spiritual self-transcendence and liberation on Earth.

Another way of looking at Yoga is to see it as a wide-ranging system of self-purification, or personal psychosomatic detoxification. This is often how traditional authorities have regarded Yoga. For instance, in his well-known *Yoga-Sutra* (Aphorisms on Yoga) the second-century sage Patanjali makes this pertinent point:

Liberation occurs when the mind is equal in purity to

*The phrase "Modern Postural Yoga" was coined by the British sociologist of religion Elizabeth de Michelis, *A History of Modern Yoga: Patanjali and Western Esotericism.* London and New York: Continuum, repr. 2006 [2004].

the ultimate Reality. (3.55)*

In other words, when our mental waters are cleansed of all polluting thoughts, all negative emotions, and all unworthy aspirations, then we can stand perfectly free, unsullied as a lotus blossom floating in a muddy pond.

Some types of Yoga, such as Raja-Yoga and Jnana-Yoga, tackle the process of self-purification via the mind itself; others, like Hatha-Yoga, extend this process to the body. The originators of Hatha-Yoga felt that much of our mental cloudiness is due to bodily impurities. In this they unknowingly echoed the wisdom of the Roman poet Juvenal, who, in the second century A.D., formulated the dictum *mens sana in corpore sano,* "a sound mind in a sound body." Juvenal, by the way, also echoed Yoga's teaching that virtue is the pathway to a tranquil life.

From a yogic perspective, self-purification is a preeminent virtuous activity, which leads to even greater virtue, which in turn leads to greater inner freedom, which spells greater happiness.

Sadly, it is no longer fashionable to talk about virtue, never mind practicing it. In this book, however, we will repeatedly invoke this old-fashioned concept, because, as will become evident, it makes implicit sense when dealing with Yoga. In speaking about virtue, we wish to indicate that Green Yoga is grounded in a reappraisal of our most cherished values, followed by a corresponding change of attitudes and behaviors.

If you are a practitioner of any of the modern versions of Hatha-Yoga, you will be no stranger to the related ideas of purity and self-purification. They are part and parcel of

---

*For a literal rendering of Patanjali's aphorisms, see Georg Feuerstein, *The Yoga-Sūtra of Patañjali: A New Translation and Commentary* (Rochester, Vt.: Inner Traditions, 2d ed. 1989).

Yoga's technical vocabulary. You will know them, for example, from the simple yogic recommendation to eat a "pure" diet or drink "pure" water to support your daily routine of physical practices, notably postures and breath control. Perhaps you also learned various cleansing techniques, such as *neti** or *dhauti,*** which are an integral part of the repertoire of traditional Hatha-Yoga.

You will likely also be familiar with the idea of cultivating mental "purity." To quote from one of the writings of B. K. S. Iyengar, who is unquestionably the best known contemporary teacher of Yoga and can be credited with having been mainly responsible for the current boom in Modern Postural Yoga:

> Yoga commences with the purification of the gross [physical] body, refines it and then moves on to clean the mind and intelligence.***

As is unmistakable from many of Shri Iyengar's talks, he believes that the purest aspect of ourselves is the transcendental Self (*atman* in Sanskrit). In our true being, we are naturally pure. How, then, is it possible for us to not recognize this purity in every moment or at all? Why are we not continuously aware

---

**Neti* denotes a technique for cleansing the nostrils by means of water or a thin string.

***Dhauti* refers to a battery of Hatha-Yoga cleansing techniques—from cleaning the teeth and tongue to cleansing the throat and stomach, etc.

***B. K. S. Iyengar, "Yoga—A Universal Culture," in *Astadala Yogamālā,* vol. 1 (New Delhi: Allied Publishers, 2000), p. 84. In fairness to Shri Iyengar, we must emphasize that the Yoga practiced and taught by him is not limited to postures but is inclusive of the higher practices and, moreover, uses the eight-limbed model of Patanjali for its theoretical underpinnings. Many of his students, unfortunately, neither share his spiritual orientation nor his passion for all of Yoga.

of our pure Essence?

The answer is quite simple: Our ordinary mind is so polluted with impurities arising from ignorance and egotism that it is like a dirty mirror. Therefore, as Shri Iyengar reiterates what many other teachers have affirmed, we must polish the mirror of the mind until it reflects things placed before it with the utmost fidelity. For him, this means purifying the body, so that the brain can function optimally as a servant of a more subtle intelligence.

Another great twentieth-century teacher and adept of Yoga was Swami Sivananda of Rishikesh; he observed:

> Life is a journey from impurity to purity. . . Self-purification is the passport to the glorious foreign land of [enlightenment].*

Now, an impure mind, which is ruled by conscious or unconscious negative emotions and urges, predictably gives rise to negative actions, that is, actions that do not benefit others. Such actions cause others physical or emotional pain either directly or indirectly.

Yoga practitioners, we assume, always do their best to minimize and, ideally, eradicate all those emotions that, like anger, envy, or jealousy, typically cause harm to others. But such negative emotions are only the tip of the iceberg. We all have numerous thoughts, emotions, and urges that do not harm others directly but do so indirectly. In fact, we are enmeshed in a web of activities that have indirect consequences for other beings and the environment as a whole. Even good people can do harm as long as they are unconscious of the effects of

*Swami Sivananda, *Sadhana* (Sivanandanagar, India: Divine Life Society, 2d ed. 1967), p. 570 and p. 630.

their actions and, as Yoga insists, also their thoughts.

It is only when we cultivate mindfulness that we can begin to bring to our awareness all those many activities that have otherwise unrecognized effects. For example, you may be very conscientious about recycling paper, cardboard, plastic, glass, cans, and so on; but you may never have given another thought to what happens with your defunct computer that ends up at the local garbage dump.

A truly mindful person would make appropriate inquiries and then would come to learn that computers are in fact enormous pollutants of the environment. Monitors, for instance, contain four to five pounds of lead, batteries contain mercury, motherboards have beryllium components, printed circuit boards contain cadmium, and steel parts are covered with an anticorrosive layer of chromium—all highly toxic materials. Add to this the cables made from plastics containing polyvinyl chlorides (PVCs) and you have a lethal mix that can seriously contaminate soil and groundwater. We need good soil to grow food and safe groundwater for drinking.

In 2004, in the United States alone, no fewer than 315 million personal computers ended up in local landfills.* Huge numbers of discarded PCs are illegally shipped to mainland China, where unsuspecting adults and children sort through the electronic parts to eke out a meager living and further contaminate their already polluted soil and rivers, not to mention seriously jeopardize their personal health. Together with television sets, cell phones, and microwave ovens, computer equipment makes for truly hazardous waste.

To give another example: You have an old PC that is still

---

*See Giles Slade, *Made To Break: Technology and Obsolescence in America* (Cambridge, Mass.: Harvard University Press, 2006). See also Elizabeth Grossman, *High Tech Trash* (Washington, D.C.: Island Press, 2006).

working well enough, but you decide to get a faster, "better" model with an elegant flat screen. You don't wish to simply discard the computer and so proudly donate it to the local school. What you may not realize or consider is that the bulky screen that comes with the old computer emits strong electromagnetic radiation that is apt to damage the eyes and possibly the brains of those students who end up using the system. A thoughtful donor would hand over a computer or, for that matter, a television set with the warning to sit at least 30 inches away from the monitor and the data processor (CPU).

Just as our own unmindfulness can be damaging to others, we ourselves are constantly victims of the unmindfulness of other people. The present disastrous state of our environment is to a large extent the combined effect of the unmindful activity on the part of all human beings.

In other words, Nature—our biosphere—is a reflection of the noosphere, that is, of humanity's collective mental or psychological state. As a hermetic maxim would have it: "As within, so without." Put differently, "The world is a mere reflection of the mind."*

For the present purpose, we would like to note that the Yoga practitioner's commitment to self-purification is adversely affected by the impurity of the biosphere and the noosphere. In other words, we must take the biological and sociocultural environment into account in our efforts to purify ourselves. To formulate things more sharply: As our biotic environment deteriorates, our efforts at physical self-purification will be increasingly hampered. Likewise, the growing "pollution" of our noetic environment will prove a mounting obstacle to our efforts at mental self-purification. It is impossible to clean the body in dirty bath water.

---

*Swami Sivananda, *op. cit.*, p. 578.

To give an example of the negative impact of environmental pollution, which no one seems to be able to escape: Traces of pesticides have been detected in the breast milk of Inuit women in areas where no pesticides are used! The World Health Organization estimates that every year some 3 million people are poisoned by pesticides, leading to 250,000 deaths.*

In a Canadian study conducted by Environmental Defence, one volunteer who had been a vegetarian all her life was stunned when she learned that trace elements of the insecticide and nerve toxin Malathion were present in her blood. All participants were found to have 18 out of 19 heavy metals (cadmium, lead, arsenic, mercury, and so forth) tested for.** And on and on. It is as if Rachel Carson had never bravely challenged the use of pesticides and written *Silent Spring* back in 1962, incurring the wrath of the chemical industry and its right-wing supporters.

Tragically, the highly toxic pesticide DDT, which Carson had singled out for censure, was banned in the United States in 1972 but is still being manufactured and shipped in large quantities to unsuspecting farmers in Third World countries.

For the past few years, we have been protesting the applicaton of Malathion for mosquito nuisance control in the small town in which we live in Southern Saskatchewan. As newcomers to the town, we were shocked to learn that this nerve toxin was used annually in such a frivolous manner. Apart from demonstrably having the opposite effect to the one intended, which is to reduce the mosquito population, fogging with Malathion has a deleterious effect on the health of residents and their

*Statistic based on a WHO news release dated September 9, 2006.

**See Environmental Defence, *Toxic Nation* (Toronto, 2005). The report is available online at www.environmentaldefence.ca. Susequent reports show similar figures.

pets, and it kills off song birds and the natural predators of mosquitoes, notably bats, frogs, and dragonflies.

Every creature, including the lowliest insect, has a place in Nature. To eliminate many species, as happens with most types of pesticide, just to get rid of one insect that is deemed a "nuisance" is, to say the least, overkill. It is simply a faulty attitude.

When we take a closer look at modern society, we will discover any number of rather unenviable features—from crime and violence to overconsumption and alcoholism, to child abuse and broken homes, to dysfunctional public institutions and avaricious corporations.* Few would disagree that societal health is at an all-time low. Little wonder that, according to a 2004 report released by the U.S. Department of Health and Social Services, the use of antidepressants by adults has almost tripled between 1988 and 2000. That the situation is similar in other nations is evident from statistics released by the World Health Organization.** Societal health, physical health, and mental health are tied together, though it might be difficult to quantify the correlations that exist between them.

How realistic is it to expect to attain individual serenity in a troubled sociocultural and natural environment? To be sure, we must at least *attempt* to achieve inner peace. Yet, as the devastation of our biotic environment leads to ever more challenging situations, we must anticipate that our sociocultural environment will also become increasingly chaotic. Crime and terrorism are bound to become more frequent and devastating. As modern society is unraveling, only someone with great

*According to the Union of International Associations, modernity is beleaguered by over 30,000 distinct global issues, which together make up our present-day planetwide crisis.

**See *The World Health Report 2006: Working Together For Health* (Geneva: World Health Organization, 2006).

inner strength will be able to avoid psychopathology. Such strength is founded in what traditionally has been called "inner purity."

At the core of any type of Yoga lies the very same process, which can be called "self-purification," or purification of oneself until even the notion of self is dissolved. As we become more and more lucid internally, the idea that we are merely a discrete human personality relaxes. At its peak, this process leads to the point where our feeling of interconnectedness with all existence eclipses the sense of being a "someone" separate from everyone else. The normally sharp boundary between "me" and "you," or "us" and "them"—subject and object—is then recognized as a needless and potentially injurious artifact.

We have given this process of inner self-purification the technical name *sattvification,* after the Sanskrit word *sattva,* which stands for the quality of lucidity and pure being. Most Yoga practitioners will have at least heard about the three qualities, or three *gunas,* of Yoga and Samkhya philosophy.

Hindu Yoga regards all of cosmic existence as an interplay between the three primary qualities (*guna*) of dynamism (*rajas*), inertia (*tamas*), and lucidity (*sattva*). Only transcendental Awareness, which is known as *cit, atman,* or *purusha,* is utterly free from these three qualities. Strictly speaking, it cannot be qualified or characterized at all. Yet, according to the testimony of mystics throughout the world, it is our true essence and, indeed, the essence of everyone and everything else as well.

The *guna* model is quite archaic, but it still has some practical usefulness. It allows us to look at all things, conditions, or situations as composites of the three psychocosmic qualities. Depending on the preponderance of either *sattva, rajas,* or *tamas,* a given physical/external or mental/internal reality will manifest the corresponding quality of brightness, luminosity, lucidity, or transparency; dynamism, changeability, flexibility, or hyperactivity; and darkness, lethargy, stability, or inertia.

From this particular yogic perspective, our modern civilization strikes us as both hyperactive and stunningly lethargic. Manifestly, it definitely lacks in clarity, transparency, and luminosity, and it is in obvious need of the kind of wisdom so plentifully available in the heritage of traditional Yoga.

This then is our launching pad for Green Yoga activism. By cultivating *sattva* in all things, we can be sure of contributing benignly to the welfare of all beings. *Sattva* brings wisdom, compassion, mindfulness, responsibility, and reverence for life. In addition, *sattva* makes us predisposed to favor reality-based knowledge, which then allows us to think and act appropriately *as well as* compassionately. The cultivation of *sattva* is a spiritual and a moral imperative.*

Green Yoga Beyond the Cave

Although it is sometimes hard to believe, we are creatures of light. As explained in *Aha! Reflections on the Meaning of Everything,* ours is a phototropic species:

> From light we come, to light we go. In between, we are both light hungry and light sensitive. We cannot live without light, and yet we also shy away from too much light—both physically and psychologically. Yet, seeing with 20/20 vision, enlightenment is a very real possibility for us individually and, given enough time,

*For a fuller discussion of the relationship between spirituality and morality, see Georg Feuerstein, *Yoga Morality* (Prescott, Ariz.: Hohm Press, 2007).

for our species as a whole.*

In view of the impending biosphere collapse, the question is whether we will have enough time. Our inner darkness—spiritual ignorance, egotism, and indifference—has created a crisis that now imperils the future of our species and all life on Earth and that, therefore, may leave humanity's spiritual potential unfulfilled.

Green Yoga calls on our species' phototropic impulse and capacity to make the urgent about-turn that alone can forestall the cataclysm that many biologists and environmentalists are predicting. It is based on the cultivation of *sattva* (the luminosity factor) and the positive values, attitudes, thoughts, feelings, and actions associated with this quality.

Green Yoga is Yoga that incorporates environmental mindfulness and activism in its spiritual orientation, especially in our time of great global crisis. It stands for a sattvic mind and a sattvic world.

In bygone ages, Yoga was often associated with the ideal of literal world abandonment. Many *yogins* and *yoginis* (female Yoga practitioners) left their home and family in search of solitude in forests and mountains. They would look for a suitable cave or a shady tree to pursue the contemplative path and live out the rest of their days apart from fellow humans.

It was the sages of the early *Upanishads* who introduced the innovative teaching of the "inner sacrifice" amounting to "inner renunciation." By contrast, mainstream Brahmanism, the age-old tradition of the priestly élite, favored all sorts of external sacrifices as the ideal way of winning the grace of one or the other deity, such as Shiva, Vishnu, or Kali.

---

*Georg Feuerstein, *Aha! Reflections on the Meaning of Everything* (Eastend, Saskatchewan: Traditional Yoga Studies, 2007), p. 8.

The great teachers of the *Upanishads* taught that what really mattered on the spiritual path was to cultivate the discipline of self-sacrifice. This meant letting go of the "ego reflex" and inwardly renouncing all one's attachments to the world and society, as well as the body and mind.

This new philosophical orientation is given beautiful expression in the *Bhagavad-Gita,* which is traditionally said to be the "cream" of the "milk" of the *Upanishads.* Supposedly composed in the fifth century B.C., this Sanskrit text of 700 stanzas puts forward the integrative teaching that the yogic process can be engaged anywhere. It does not require a remote cave or dense forest for its success.

On the contrary, the enlightened adept Krishna, who is credited with this teaching, makes the point that often this kind of extreme practice does not lead to inner peace, never mind enlightenment. In his own words:

> One does not get to enjoy *action transcendence** by abstaining from actions, nor does one approach perfection by renunciation alone.
>
> For, not even for a moment can anyone ever remain without performing actions. Everyone is unwittingly made to act by the cosmic qualities**.

---

*The difficult term *naishkarmya,* here translated as "action transcendence," denotes freedom from the binding effects of one's action. There are no such effects, because the practitioner of Karma-Yoga (Yoga of Action) engages action in a selfless manner and as service. Such action lacks the ego as reference point, and the ego is a necessary precondition for action to have karmic repercussions.

**The cosmic qualities, which have already been discussed briefly, are *sattva, rajas,* and *tamas*. These describe patterns in both macrocosmic and microcosmic (psychological) phenomena.

> Someone who restrains the organs of action* but sits remembering with the mind the sense objects is called a befuddled hypocrite.
>
> More excellent than that is someone who, controlling the senses with the mind, embarks unattached on the Yoga of Action with the organs of action.
>
> You must do all allotted actions, for action is superior to inaction; not even your body's processes can be accomplished by inaction.
>
> This world is bound by action, except when this action is sacrifice. With that intent, engage in action free from attachment.**

Only self-transcending action is sacrificial, that is, a selfless offering, which contains no element of attachment. Because it does not involve the ego, it also does not cause suffering arising from the ego. Put into yogic language, self-transcending action has no karmic consequences.

Now, it is very important to know that acting selflessly is not the only criterion in Karma-Yoga. For an activity to be self transcending, it also must be the *right* kind of action, that is, it must be morally sound and appropriate.

Otherwise, one might argue that a homicidal maniac who kills people in order to "liberate" them from their apparent

---

*The five organs of action comprise the mouth, hands, feet, gentials, and anus. These are contrasted with the five cognitive organs, which are the five senses.

***Bhagavad-Gita* 3.4-3.9. The present paraphrase is an adaptation from a complete and literal translation of this text done by Georg Feuerstein, which will be published as part of a forthcoming larger work.

suffering would qualify as a practitioner of Karma-Yoga. In his own eyes, he would render them a selfless service. In the eyes of the rest of the world, however, his actions would be absurd and morally reprehensible. Of course, his presumed selflessness would be little more than the grandiose delusion of an out-of-control egotist.

Activity that is morally sound, however, is not inevitably identical with consensus morality. The latter is not always sound or valid. This can readily be seen when we consider the changing fashions in sexual mores. More concretely speaking, the separation of Church and State and the rise of secularism have led to the widespread moral relativism we can witness today, which is epitomized in the popular hedonistic saying "If it feels good, do it!"

The best guide in moral matters is not any formulaic catechism on ethics but wisdom itself. And wisdom, as indicated previously, flourishes through the nurturing of the quality of *sattva* in our mind, behavior, and surroundings.

In our present era, sound action must include morally sound societal and environmental *activism*. What this means will be made clear on the following pages. The *Bhagavad-Gita,* which Mahatma Gandhi called "the universal mother,"* is fundamental to the spiritual activism of Green Yoga. We will refer to it and its Yoga of Action on many more occasions.

Yoga is essentially a liberation teaching. This means that it aims at freeing a person from the illusion of being a distinct, encapsulated entity that is separate from all others. In the course of its long history, Yoga has adopted diverse, even contrasting, philosophical ideas. Whatever its metaphysical undergirding might be, Yoga affirms that the ego-personality that we normally animate is illusory, or fictional.

*Cited from Gandhi's article in *Harijan*, August 24, 1934.

What, you will ask (if you are new to this idea), "I am not real?" Yes and no. Contrary to our most inveterate belief, we are not a particular, discrete body-mind. Rather, our true essence is the true essence of everyone and everything. And what is that? It is the whole of reality; it is the Whole. Another way of talking about the Whole, which is ultimately real, is to say that everything is interrelated and interdependent. This is a central concept of the Yoga tradition.

Liberation, which is also frequently referred to as the condition of enlightenment, is the case the moment we stop identifying with our historical body-mind and instead identify with the Whole itself. This is not merely a belief or thought. Beliefs can change and thoughts do pass. Rather, liberation is a realization that concerns our entire being. If it is limited to our head or mind only, it is not actual liberation.

Sometimes a distinction is made between ultimate liberation and enlightenment. In this case, the former is deemed a transcendental state of perfect freedom, which is beyond space-time and embodiment, whereas the latter is regarded as inner freedom while yet embodied. Strictly speaking, however, freedom is unqualified; hence we can use both concepts equivalently.

The Two Ethical Pillars of Green Yoga

Green Yoga can be viewed as a modern continuation of the spiritual activism transmitted in the *Bhagavad-Gita.* It is concerned not just with personal salvation but with the welfare of all species and the Earth as a whole.

The core social ideal of Karma-Yoga is *loka-samgraha.*

This Sanskrit phrase has been translated in many different ways. The term *loka* means "world" or "people" and presents no big problem to the translator. This is not the case with *samgraha,* which has variously been rendered as "integration," "maintenance," "control," "cooperation," "solidarity," "guidance," and "welfare."

However we may wish to reproduce this term in English, loka-samgraha is one of the two ethical pillars of Green Yoga. Unfortunately, the *Gita* does not explain the concept of loka-samgraha and also the classical commentaries on this work offer no helpful elaborations. Remarkably, this concept apparently does not occur anywhere else in the voluminous Sanskrit philosophical literature. We can only gauge its meaning from the context in which it occurs in the *Gita.* If our interpretation is correct, loka-samgraha stands for the ideal of working for the material, social, moral, and spiritual benefit of the whole world.

The second pillar of Green Yoga is the Buddhist *bodhisattva* ideal. Like loka-samgraha, it represents the kind of yogic process that we have called *sattvification,* though we realize that Buddhism does not subscribe to the *guna* model. Still, Buddhist Yoga also understands the path to liberation as one of purification and thus increasing illumination in the broadest sense of the term. According to H. H. the Dalai Lama, the bodhisattva ideal can be characterized as follows:

> What do we mean by *Bodhisattva*? *Bodhi* means enlightenment, the state devoid of all defects and endowed with all good qualities. *Sattva* refers to someone who has courage and confidence and who strives to attain enlightenment for the sake of all beings.*

*Dalai Lama, *A Flash of Lightning In the Dark of Night: A Guide to the Bodhisattva's Way of Life* (Boston, Mass.: Shambhala Publications, 1994), p. 12.

The wish to reach enlightenment for the sake of others, which is known as *bodhicitta,* must be crisp and clear. A mere vague sentiment of wanting to help others will not do. For bodhicitta to be genuine, it must be grounded in real understanding of the nature of enlightenment and also of the nature of empirical existence.

A bodhisattva has a keen sense that all beings are suffering in one way or another and that wisdom is the sole means of overcoming this constitutional suffering. This is not a matter of mere ideology but a self-evident truth, which Gautama the Buddha discovered 2500 years ago. We only need to examine our own life to recognize that we are subject to all sorts of discomfort, sickness, pain, frustration, disappointment, anguish, and fear, and that happiness is rather elusive.

Much of our life is dedicated to the pursuit of happiness, though by and large we hunt for it in the wrong places, and we typically end up with mere transient pleasure followed by renewed suffering. Even when we have recognized that this is so, we still feel compelled to seek pleasure rather than abiding happiness and inner peace.

The problem is that our misguided quest revolves around ourselves, our personal happiness. The bodhisattva ideal, by contrast, places the happiness of others before our own and thus gives us the key to inner peace. For paradoxically, when we start being more concerned about the ultimate welfare of others, the ego goes out of focus. As long as we focus on the ego, our vision of others becomes blurred.

Manifestly, the loka-samgraha ideal of Karma-Yoga and the bodhisattva ideal of Buddhism have much in common. In our opinion, however, they are not quite identical but emphasize aspects of the same caring concern for others. If, as we propose, the loka-samgraha ideal is conceived as welfare in a broad sense, the bodhisattva ideal has, by comparison, primarily a spiritual thrust. It involves the deeply felt wish to help others achieve liberation, so that their suffering comes to an end. If material or emotional support can bring others closer to their own liberation, then the bodhisattva will prove a generous and patient friend.

The bodhisattva commits to striving for liberation with all his or her might, so that he or she can be of maximum spiritual benefit to all other beings and alleviate their condition of suffering. The bodhisattva wastes absolutely no time and regards the enlightenment of others as a most urgent task. To this end, the bodhisattva vigorously cultivates all manner of skillful means based on wisdom—from patience and generosity to kindness and compassion.

Striving for enlightenment, or liberation, and working for the welfare of others are manifestly not mutually exclusive. On the contrary, from an integral perspective, they are indeed complementary.

In his marvelous Sanskrit classic *The Way of the Bodhisattva* or *Bodhicaryavatara* (1.12), the eighth-century Buddhist scholar and abbot Shantideva wrote:

> Virtues, like the plantain tree,
> produce their fruit,
> but then their energy is spent.
> Only the wondrous tree of *bodhicitta*
> bears fruit and then continues to grow.*

---

*This is a paraphrase only. For a translation of the most significant verses of Shantideva's text, see Georg Feuerstein, *Epitome of Shantideva's "Entering the Path to Enlightenment" (Bodhicaryavatara-Samgraha)* at www.traditionalyogastudies.com/articles_translations_bodhisamgraha.html.

2
All OUR MOTHERS

One Cosmic Womb

The peoples of the Stone Age knew that despite all their differences, they were the children of the same Universal Mother. Today, cosmologists tell us that the cosmos began with the Big Bang—a violent event thought to have occurred some 15 billion years ago. This scientific creation "story," which is really what this model amounts to, reminds us of the same essential fact: All things emerged out of the same original Singularity, and thus we and all other beings, as well as the stars and planets, are all blood relations to one another.

Cosmologist Brian Swimme, a professor at the California Institute of Integral Studies in San Francisco, put it this way:

> *The birthplace of the universe, where existence first sprang forth, is fifteen billion light-years from the Earth.*
>
> Most physicists regard this discovery of the birthplace of the universe as the most significant of the twentieth century. . .*

*Brian Swimme, *The Hidden Heart of the Cosmos* (Maryknoll, N.Y.: Orbis, 1996), p. 2. (Italics are in the original.) Newer models favor 13.7 billion years for the life span of the universe.

Why is it so important for us to encounter—through photons travelling in space since the beginning of time—the origin of the world we live in? As Swimme sees it, this discovery gives us the opportunity to unify humankind through the convergence of scientific knowledge and ancient wisdom. Both traditions lead us to a similar understanding and a sense of awe-inspiring mystery.

To know that there has been a single beginning for all of existence also means to know that whatever our individual trajectory might be, it must be part of the trajectory, or destiny, of the entire universe. Furthermore, the same basic "stuff" that created billions of galaxies and continues to go into the making of new stars also can be found in our body. We are, as astronomer Carl Sagan noted in his bestselling book *Cosmos*, "starstuff pondering the stars."*

Science cannot and does not tell us what caused the Big Bang or what existed before this event. Such questions are the province of metaphysics, theology, and religion. But science has brought us to the edge of what is knowable. Through technological inventions like the microscope and telescope, science has revealed to us the breathtaking grandeur of the cosmos. Inadvertently, it also has given us incontrovertible proof of the interconnectedness of everything.

Thus, the interdisciplinary science of ecology, which is a branch of biology, studies the interaction between life forms and also between life forms and their abiotic environment. Ecology is built on the recognition that life on our planet is not a haphazard affair. Rather, life is a great symphony in which all the numerous life forms have their particular role to play. Of course, ecology has formulated its own technical vocabulary for talking about this. It speaks of "biodiversity,"

---

*Carl Sagan, *Cosmos* (New York: Ballantine, 1985), p.300.

"biome," "biosphere," "ecosystem," "food chain," "habitat," "niches," and so forth.

The poetic metaphor of the Web of Life describes well the systems approach of ecology.* The language of ecological interconnectedness and interdependence reached its apotheosis in the 1960s when biologist James Lovelock gave our Earth's biosphere the ancient Greek name *Gaia*.** He wanted to acknowledge with this term that the biosphere is a finely tuned system that has its own systems logic. Although he did not mean to suggest that the Earth itself was an intelligent entity, this is exactly what it came to mean for countless people—Mother Earth. The idea itself has a long history, as do the various sentiments that come along with it.

As a scientific discipline, ecology tries to stay clear of theological explanations and aligns itself with systems-theoretical thinking and empirical facts. Although this science is still in its infancy as it were, it has made some very important findings and also has given us extremely valuable insights into how our planet functions and how humanity is interfering with Earth's natural balance.

Interconnectedness, equally importantly, also is pivotal to quantum theory. Once electrons have come into contact, they will always remain in contact, however far apart they might be spatially. This notion was an early feature of quantum theory, but then was mostly ignored until the Irish physicist John S. Bell formulated in 1964 the well-known theorem named after him, which has since been experimentally verified.

This quantum connection happens, it would appear, at

---

*See Fritjof Capra, *The Web of Life: A New Scientific Understanding of Living Systems* (New York: Anchor Books, 1996).

**See James Lovelock, *Gaia and the Theory of the Living Planet* (London: Gaia Books/Octopus, 2005)).

a speed faster than the speed of light. According to relativity theory, this is an impossibility; nothing can exceed the speed of light. Thus far, however, no convincing alternative explanation has been found. The superluminal handshake at the quantum level is an experimental fact, which is known as "entanglement."

Quantum theory, although very successful in explaining quantum phenomena, is a veritable philosophical quagmire. It clashes fiercely with Einstein's relativity theory. The former theory insists that quantum reality is in fact nonlocal, a mere statistical probability. Einstein balked at this notion, and his relativity theory strictly affirms that reality is somehow "out there," that is, local.

Both relativity theory and quantum theory certainly have put us in touch with some vexing paradoxes, which drive home the point that the universe, unlike human beings, does not readily lend itself to being squashed into conceptual boxes. In any case, the latter theory has also presented us with yet another demonstration of deep interconnectedness.

To summarize, astrophysics, quantum physics, and ecology—each discipline in its own way communicates to us the same fundamental truth, namely that nothing that exists is truly independent or isolated from everything else. As John Muir wrote, "When we try to pick out anything by itself, we find it hitched to everything else in the Universe."*

The ancients knew this very well. But ever since the European Renaissance, which introduced a new mind-set, we have believed that the human being is something very special and apart from, or superior to, Nature. From the Renaissance to the so-called Age of Enlightenment and the scientific endeavor

*John Muir, *My First Summer in the Sierra* (Boston, Mass.: Houghton Mifflin, 1911), p. 110.

of controlling and dominating Nature through technology was a comparatively small step. We already know the long-term outcome of this misguided worldview: the present global crisis, which in due course will make it very clear who or what is in charge.

The World Process: Food For Food

The idea of the interdependence of all things was first articulated in the *Rig-Veda,* which is the oldest known "scripture" in any Indo-European language. Estimated to date back five millennia ago, this work, composed in archaic Sanskrit, comprises 1,028 hymns containing 10,600 verses.* Most of these are invocations to deities, but some have a philosophical and metaphysical flavor.

In the present context, one hymn (10.90) in particular strikes us as relevant. This is the so-called Hymn to Man, or Purusha-Sukta, which poetically looks upon the entire universe as issuing forth from Primordial Man, the Macranthropos, who stands for the original Singularity mentioned earlier. Even though, at the beginning of time, Macranthropos poured himself forth in the form of the universe, he yet transcends it.

A similar conception can be found in the *Atharva-Veda,* which may be of a slightly younger age than the *Rig-Veda.* In one hymn (10.8), the ultimate Singularity is mythologically

*Scholars disagree about the date of the *Rig-Veda.* Those who subscribe to the questionable Aryan Invasion hypothesis formulated in the nineteenth century tend to place this hymnody between 1200 and 1800 B.C. This seems far too late, however. See Georg Feuerstein et al., *In Search of the Cradle of Civilization* (Wheaton, Ill.: Quest Books, 1998).

presented as the Creator Prajapati, who fashioned the universe and yet transcends it. He, who is the everlasting Whole (*purna*), is most eminently visible as the Sun, which is indeed literally the creator and sustainer of our local "universe," the solar system. Long before the Sun, a middle-aged star, finally ceases to give off life-bestowing light and heat, which astrophysicists tell us won't happen for another 5 billion years or so, life throughout the solar system will have become extinguished.

This Vedic hymn also speaks of "the stretched-out thread" that strings together all creatures, and the thread itself is said to abide in the ultimate Singularity, which is the thread of all threads, the incommensurable superstring. Inspired by this archaic notion, some later Vedanta philosophers introduced the concept of the "Thread-Self" (*sutra-atman*).* As they explained, this Thread-Self is the totality of *subtle bodies* that constitute the "soul" of all living beings. The life principle, as they put it, is like thread pervading a piece of cloth. It animates all living beings endowing them with the capacity for sentience and action.

The Thread of Creation is nothing other than what we have referred to as the Web of Life. It reflects, at the level of the manifest universe, the flawless unity that marks the transcendental Reality, or Singularity, as such. That is to say, what appears as Singularity at the transcendental level appears as connectedness and interdependence at the empirical level. Metaphysics and theology deal with the former, while ecology deals with the latter. Quantum theory curiously straddles both levels, which is undoubtedly why it has won great popularity outside the domain of physics, even though few laypersons can

*See, e.g., Vidyaranya's *Panca-Dashi* (6.200). The Thread-Self is also known as the Golden Germ (*hiranya-garbha*).

claim to comprehend it. We do not make such a claim either.

Even if quantum theory should turn out to be wrong or only partially correct, experiments have overwhelmingly demonstrated a remarkable interdependence at the subatomic level. This interdependence is mirrored at all other levels of existence, be it biotic, societal, psychic, or cosmic.

During the era of the *Upanishads,* which followed the Vedic age, the idea of connectedness was expressed in what one could call "food mysticism." Thus, in the three-thousand-year-old *Taittiriya-Upanishad* (2.2), we can read:

> From food, verily, whatever creatures dwell on Earth are created. By means of food alone they live and, moreover, into it they finally pass. For truly, food is foremost among things. Hence it is called a panacea. Verily, those who worship the Ultimate (*brahman*) as food surely obtain all food. For truly, food is foremost among things. Hence it is called a panacea. From food all beings are born and by means of food they grow. [All] beings eat and are eaten. Hence they are called food (*anna*).*

This amazing passage talks about the ecological notion of the Food Chain. It makes the point that, at the material level of existence, all living beings are food for each other: Big fish devour little fish, and little fish eat even smaller fish.

To sustain the human body, we must likewise consume other life forms, be it plants or animals. Most humans opt for an omnivorous diet, but there are also hundreds of millions who, because of customs, religious doctrines, moral considerations, or health concerns have adopted a vegetarian diet. Relatively

*Translation by Georg Feuerstein.

few people are vegans, shunning meat and fish, as well as all animal products (such as milk, cheese, eggs, honey, fur, leather, wool, silk). Vegans usually subscribe to a strong philosophy, which has nonharming at its core. We will return to this subject shortly.

Here we must appreciate that even plants are a form of life. Plants may not have consciousness, but, as experiments have shown, they are sensitive not only to light but also to other influences. This sensitivity was demonstrated by the Bengali physicist and biophysicist Jagdish Chandra Bose (1858–1937). By means of a crescograph, a device he had invented for measuring plant growth, he demonstrated that plants possess the equivalent of a nervous system. They would grow faster to pleasant music but withered when exposed to discordant noise. After analyzing the electrical membrane potential of plants, Bose came to believe that plants could experience pain and were even responsive to affection. His findings have been confirmed by other researchers. This gives loving broccoli or spinach a whole new meaning. Seriously, though, if plants experience pain, what are we to do?

In order to avoid causing harm, do we then have to stop eating altogether? Yoga's answer would be that since we are alive, we must do everything necessary to maintain our physical existence. At the same time, however, we must *minimize* harming other beings. To starve ourselves to death is not considered an acceptable solution.*

Besides, perfect nonharming is anyway impossible in this rather imperfect realm. Even a breatharian is, strictly speaking, not exempt from taking life. For with every gulp

*The practice of voluntary death (*sallekhana*) by extreme fasting is condoned in Jainism only for ascetics who are seriously sick beyond any hope of recovery and are capable of exiting the body consciously.

of air, we gobble up countless microscopic beings. This is a very sobering realization, which only Jaina monks have taken absolutely seriously. Their code of ethics, fashioned around the time of the Buddha, demands that they make every effort to avoid killing even the most miniscule life forms. They filter the water they drink, wear a piece of cloth over the mouth to avoid inhaling insects, and clear the path on which they walk, so that they do not inadvertently step on beetles or ants. Even these extreme precautions, however, are not enough to rule out mayhem at the microbic level.

The human body is teeming with microscopic life. In fact, our body serves as an entire planet to literally hundreds of trillions of microbes, something like two pounds worth. There are twenty times more microbes in the human body than it has cells (estimated between 10 and 50 trillion). Thus, the number of body-resident microbes surpasses the number of galaxies in the known universe by a factor of at least 100,000.

Obviously, some microbes are detrimental to our physical well-being. These are the ones responsible for the flu, infections, tuberculosis, AIDS, and so forth. But then again, some microbes are necessary for the maintenance of our health. Thus, the bacteria in our intestines help break down the food we eat. Others live on our skin and in our mouth, armpits, genitals and so forth, where they have a variety of useful functions. As long as this symbiotic relationship remains balanced, all is well. However, when the host of bacteria resident in our body grows too strong and demanding, our body suffers.

It is easy enough to draw a parallel between this and our own relationship to the planet at large. Six billion people living mostly in an unsustainable way is a threat to the planetary body we call Earth. The Earth weighs roughly 6 billion trillion tons, and assuming an average weight of 35 kilogram per person (young and adult, big and small), humanity weighs in at around 21 million tons. This yields a proportion of c. 1:300 trillion, which is not altogether dissimilar to the proportion that pertains between the weight of a single microbe and our body. Even though humanity's total weight is insignificant by comparison with the Earth's weight, the impact we have is disproportionally weighty. More accurately, our impact on this planet has become supremely destructive. Our species behaves like runaway microbes.

Just as we devour life in order to stay alive, once we succumb to death, unless we opt for cremation, our body in turn is consumed by living creatures in the form of microbes.

The Upanishadic passage quoted earlier makes another profound point, which is easy to overlook when our focus is fixed on material explanations. The anonymous sage who uttered those words also referred to the "worship of the Ultimate" as a sure means of securing "all food." How are we to understand this?

There are two explanations. *Exoterically*, this statement suggests that whenever we attune ourselves spiritually to the Source of all life, to the ultimate Singularity, we can expect an effect that in physics is known as "entrainment." This is when oscillating systems with different periodicity fall into synchrony much like a battalion of soldiers marches perfectly in step.

Colloquially, this phenomenon is hinted at when we say "Go with the flow!" The suggestion implies that we should surrender our egocentric struggle to allow our larger innate intelligence to step in and guide our behavior.

Esoterically, again, the Upanishadic statement is a statement about mystical union, or ecstatic self-transcendence. In the ecstatic state, which is known in all the religio-spiritual traditions of the world, we experience directly the Web of Life. We experience with breathtaking immediacy that the entire universe is a perfect Whole in which every being and every thing is interconnected.

In ecstasy, our consciousness is as it were spread out over the entire cosmos. We become the Macranthropos, the Cosmic Man. Consciousness, as manifesting in ecstasy, is the "thread" that runs through everything. In the already cited *Taittiriya-Upanishad* (3.10.6–7), the anonymous sage who was probably also responsible for the rest of the passage exclaims:

> Oh, wonderful! Oh, wonderful! Oh, wonderful!
> I am Food! I am Food! I am Food!
> I am the Food-Eater! I am the Food-Eater! I am the Food-Eater!
> I am the Maker of Poetry (*shloka*)! I am the Maker of Poetry! I am the Maker of Poetry!
> I am the first-born of the Cosmic Order (*rita*), prior to the deities, [residing] in the hub of immortality!
> He who gives Me away [as food], he indeed preserves Me!
> I, who am Food, eat the Eater of Food!
> I have overcome the whole world!
> [My] effulgence is like the Sun.*

The universe of multiple cross-links and interdependences is mythopoeically both food and food eater, depending on one's perspective (i.e., depending on whether one consumes or is consumed). Uplifted into a state of mystical union, the inspired

*Translation by Georg Feuerstein. The phrase *shloka-krit,* here rendered as "maker of poetry," suggests that the sage absorbed in ecstasy experiences himself as the transcendental origin of the secret proclamation (*upanishad*).

sage abides as the transcendental Witness, or Experiencer, of absolutely everything. As such, he is universally present as both "food" and "food eater."

As long as we experience ourselves as finite beings—humans or nonhumans—we either are food for someone else or are consuming others. In the condition of cosmic consciousness, by contrast, we experience ourselves simultaneously as all food and all food eaters.

In the absence of such mystical elevation, we have good reason for cultivating humility, because in the end every being is food for someone else. Even the most voracious creature will sooner or later succumb to death and at least at that point become a tasty morsel for others.

A related consideration: Even at the level of our thoughts, we are both food and food eaters. We imbibe others' ideas and, in turn, our ideas become "food for thought" to others. That is to say, the interdependence we witness in the biosphere also is the case in the noosphere (the realm of ideas).

From the perspective of Yoga, the greatest adventure on which we could possibly embark in life is the quest for liberation. It is through this quest that we can discover first hand the transcendental Thread of Life, the "Thread-Self."

This discovery is not a mere intellectual insight. It is a shift in our whole being, for through the state of ecstasy, we discover our identity with that ultimate Being that is the hub of all existence. The Thread-Self is our true nature, which, when we rediscover it, dispels the powerful illusion that we are an island unto ourselves, separate from everyone and everything else. We *become,* in consciousness, the Thread of Life itself.

This is the gist of the experience of cosmic consciousness, or, in Patanjali's terminology, samprajnata-samadhi, the ecstatic identification with a given object, from a mere atom to the universe at large. We can consciously and ecstatically identify with everything because we are, in reality, everything.

Reverence for All Life

As we have seen, one of the pivotal ideas of Yoga is that we become what we contemplate with sufficient vigor or persistence. In other words, we shape our reality depending on the energy we invest in a given thing. And what we invest our energy in depends largely on our viewpoint and our values.

If our values are shaped by the overwhelming belief that we inhabit a one-dimensional, material world bereft of spiritual depth, we will behave accordingly. We will eat, drink and be merry, and we will eventually die, fully expecting a void to engulf us at the moment of death. From such a materialistic viewpoint, there would indeed be very little reason to worry about the hereafter or the long-term consequences of our actions on Earth.

Yoga offers a potent antidote to this impoverished view, which tends to spawn egocentric values and attitudes. There are of course morally upright and also caring individuals among those who have embraced a materialistic philosophy. However, their materialistic, secular humanist ideology, will predictably impede their learning about the full potential of humanity, which includes the possibility of spiritual growth. It obliges them, for instance, to offhandedly dismiss valuable ideas like liberation, transcendence, karma, and reincarnation, which are fundamental to any truly comprehensive spiritual ethics.

Human beings have a remarkable chameleon-like capacity

for "make-believe." That is to say, if our mind assumes that we exist in a purely material universe, we are apt to behave as purely material (biological) creatures. If, however, we affirm that the universe is very much larger than the material world, we tend to behave as creatures capable of great self-transformation through which we can then discover the hidden depth of existence, the splendid dimension of Spirit, or *ultimate* interconnectedness.

We all make basic assumptions about existence. Even skepticism, cynicism, nihilism, and "I-don't-care-ism" amount to a quasi-philosophical stance. They are ad hoc "theories" based on shallow values and naive notions about life. Why this matters is because our basic assumptions about life are characteristically expressed in our attitudes and behaviors, especially in our relationship to animals and to Nature as a whole.

If we are in the habit of considering only the material level of existence, we tend to see animals (and often even other human beings) as exploitable objects and Nature as an inexhaustible resource intended for our free use. We propose that it is precisely this attitude that is largely responsible for the current environmental crisis.

Speaking of a truly ecological lifestyle, which is based on sound philosophical principles, Arne Naess observed:

> Never use life-forms merely as means. Remain conscious of their intrinsic value and dignity even when using them as resources.*

Our lack of reverence for all life is particularly apparent in the ruthlessness with which we exploit the animal kingdom for

---

*Arne Naess, "The Place of Joy in a World of Fact," in *Deep Ecology for the Twenty-First Century,* ed. by George Sessions (Boston, Mass: Shambhala Publications, 1995), p. 360.

food and for a large variety of consumer goods. Here we wish to focus on people's taste for meat and fish, as well as animal-based food products, which has all sorts of major environmental consequences that are seldom fully understood.

The annual global meat production is nearly 300 million tons, while the total milk output is over 600 million tons. Our voracious appetite extends to fish. Every year, over 100 million tons of fish are harvested from the oceans. According to a statistic issued by the international journal *Nature* (May 15, 2003), 90 percent of all large fish in the ocean—from swordfish, marlin, sharks, tuna and so forth—are gone. In other words, we have been severely overfishing and over-consuming. The human-made toxins, notably mercury, in the ocean and other bodies of water may also play a role in this unbelievable decimation of large fish.

Now, some people might argue that all this is to be expected with 6 billion mouths to feed. Exactly! But we can't leave it at that, thinking that there is absolutely no problem here. Only the most uninformed still believe that 6 billion people can live sustainably on this planet.

It is true enough that if the developed nations were to stop overconsuming and share their plenty, the poorer countries of the world would not have so many millions of people literally starving to death and another 850 million or so suffering from chronic hunger every year. However, even a greater equity between the rich nations and the so-called Third World would not be a long-term solution to the environmental overload our Earth is experiencing.

The world population continues to grow by about 90 million people every year, which is more than the population of Vietnam, Germany, or Egypt. At the present rate of growth, the human family is predicted to number over 8 billion by the year 2030. This is highly unlikely to happen, however, because environmental devastation is coupled to population growth and increasing consumption, finally causing a collapse of the biosphere.*

China, with its 1.3 billion citizens, is the first nation to recognize the population problem and also to do something about it. Some of China's measures to control population growth—such as enforced abortions—are clearly reprehensible. Nevertheless, we must expect that other nations, including Western democracies, will before long be forced to adopt draconian measures as well. Those measures will undoubtedly cut deeply into some hard-won democratic freedoms.

It stands to reason to think that as environmental conditions become more and more degraded, food production will be seriously affected. As we will show in the next chapter, the availability of potable water is also becoming an increasingly critical problem. **To avoid complete biosphere collapse, the human family will have to voluntarily control population growth.** Understandably, the political leaders of the so-called "free world" are reluctant to regulate people's reproductive rights, but realistically they will sooner or later have to follow China's lead.

Overpopulation is evidently an enormous problem. But so are overconsumption and consumption higher up in the food chain. This brings us back to the consideration of the exploitation of the animal kingdom, meat eating, and vegetarianism.

Although this topic is rather unpopular even in avowed spiritual circles, we must pay proper attention to it, because when people enjoy a hamburger or a beefsteak, which includes

---

*See Donella Meadows, Jorgan Randers, and Dennis Meadows, *Limits to Growth: The 30-Year Update* (White River Junction, Vt.: Chelsea Green, 2004).

many Yoga practitioners, they are directly contributing to environmental degradation. Here is how: It takes roughly 40 million tons of plant protein to feed America's c. 100 million heads of cattle. Yet, this translates into only 7 million tons of animal protein. The United States alone could feed 800 million people from the grain that its livestock consumes.

Consider this: meat eaters require at least seven times more land than vegetarians! Furthermore, it takes 13,000 liters of water for each kilogram of beef. We will talk about the growing shortage of water in Chapter 3.

According to a report released in 2006 by the United Nations Food and Agriculture Organization, livestock around the world is responsible for 18 percent of greenhouse gas emissions, which tops the emissions from automobiles. To be precise, livestock generates 65 percent of human-related (anthropogenic) nitrous oxide, which is said to have 296 times the global warming potential of carbon dioxide (CO_2); 37 percent of all human-related methane, and 64 percent of ammonia (a major cause of acid rain).

In other words, livestock contributes significantly to the Greenhouse Effect, which was first identified as long ago as 1824 and whose existence has finally been officially acknowledged by governments. Our planet is indeed heating up and giving us rising sea levels, megahurricanes, unpredictable climate, and much else.

Moreover, cattle-grazing degrades and erodes the land. Crops grown for livestock are sprayed with pesticides and thus pollute the rivers and ultimately the ocean, which already have high levels of pollution. Governments and private institutions have started to have bans and advisories on fish consumption, because some species show significant levels of toxic chemicals as a result of pollution by the livestock industry and other industries.

It would appear that hog "factories"—of which there are over 150,000 in the United States alone—are the worst polluters in the meat industry, polluting land, air, and water. To mention just one of many disasters, in 1995 an eight-acre hog-waste lagoon in North Caroline spilled 25 million gallons of manure into the New River. This spill killed an estimated 10 million fish. Four years later, when hurricane Floyd pummeled the same state, another five hog-waste lagoons burst and 47 were flooded, polluting land and water.

In 2001, the United States' Environmental Protection Agency forced five hog factory farms in Oklahoma to provide residents with free bottled water, because their operations had badly contaminated the local drinking water. Alas, we now know that bottled water is not necessarily much better than the treated water flowing from a household tap; it can even be worse.

As for air pollution, anyone living downwind from a hog farm can readily testify to the abominable odor. Hogs produce an enormous amount of manure, which translates into the hazardous chemicals methane, ammonia, carbon dioxide, and hydrogen sulfide. The last-mentioned gas (H_2S) has a particularly adverse effect on human health and with intensive or continued exposure can even lead to severe brain damage and death.

There are an estimated one billion heads of hog worldwide, with China claiming the lion share of over 400 million and the United States as a distant second with about 60 million. Around the world, pork is clearly people's meat of choice. Unfortunately, this dietary preference comes with a huge ecological price tag of which the consumer is generally unaware. *Caveat emptor!*

Much more could and in fact has been said about the dire

environmental consequences of meat consumption.*

Much has also been written about the negative impact of meat eating on our physical well-being. Meat consumption has been related to cardiovascular diseases, cancer, diabetes, and osteoporosis.**

In its bid for bigger profits, the livestock industry has adop-ted practices that represent a serious health challenge not only to the animals but also to meat eaters. Animals are routinely treated with antibiotics and growth hormones. In the United States, approximately 1.2 million tonnes of antibiotics are added to animal feed every year. Apart from promoting ill health instead of health in livestock, these substances end up in the body of meat eaters, where they contribute to hormone imbalances, interfere with the reproductive system, and may cause a variety of cancers.

If the ranks of vegetarians in Western countries are swelling, it is perhaps because people are beginning to realize that the meat they eat has become unsafe for human consumption.

In addition to environmental and health concerns, we must also look at the ethical issue of the immense cruelty commit-ted against animals by the meat industry, especially in "fac-tory farming." Calves are forced to eat a monotonous milk-replacing diet, and cattle are fed unmarketable parts of their own kind. Chickens are debeaked (without anaesthetics) and raised in torturously small wire cages where movement is impossible. Young cocks are considered worthless and are

*See, e.g., John Robbins, *A New Diet for America* (Tiberon, Calif.: H. J. Kramer, repr. 1998).

**See, e.g. William Harris, *The Scientific Basis of Vegetarianism* (Honolulu: Hawaii Health Publishers, 1998). Excerpts from this book can be found online at www.vegsource.com/harris/book_contents.htm.

either killed with carbon dioxide or they are simply shredded alive. Ducks and geese are treated even worse. Pigs are taildocked and kept in miniscule pens in the dark, and so on. The final slaughter of these animals is a typically brutal affair. As Dirck Van Sickle wrote in his still readable 1971 book *The Ecological Citizen*:

> There is a fundamental immorality in raising pigs, or any animal, in conditions which are frequently so intolerable that only heavy drugging suspends sudden dieback. All life should at least have the chance, as much as its consciousness will allow, to experience sunlight and space to move, if not other joys of simply being alive.*

Anyone who has ever watched one of the PETA documentaries on the kind of cruelty to animals committed on a large scale by the meat industry will appreciate the magnitude of this problem. Our (karmic) debt to the animal kingdom is massive. We can only recommend that you oblige yourself to watch at least the documentary "Meet Your Meat" featured at www.meat.org.

We warn you, this will not be a pleasant experience! On the contrary, it will probably make you gag or cry. But this is the somber reality behind the neatly packaged meat displayed on supermarket shelves and the nicely garnished meat dishes prepared in restaurants.

Yoga practitioners, and really everyone else as well, need to know how the meat and other food they consume is produced. If Yoga is about the transcendental Reality, or Spirit, it is also

*Dirck Van Sickle, *The Ecological Citizen* (New York: Harper Perennial, 1971), p. 72.

about the reality of our everyday life.

Yoga practitioners like to talk about wholeness, health, and harmony. None of these three ideals are present in a lifestyle of overconsumption and inconsiderate meat eating. We both have enjoyed a vegetarian diet for years, and we both feel that even if a non-meat diet were to be detrimental to our health, we would not consume meat for all the reasons mentioned above.

For centuries, literally hundreds of millions of people have thrived on a vegetarian diet, and there is no need to burden our stomach and intestinal tract with having to break down meat products. For many years, the personal physicians of H. H. the Dalai Lama, had recommended to him a meat diet, even though he himself has long preferred a vegetarian cuisine. On April 5, 2005, the Dalai Lama publicly announced that the kitchen at his residence in Dharamsala, India, had switched to a fully vegetarian diet. Since then he has strongly urged the Tibetan Buddhist community to follow suit.*

H. H. the Karmapa, the spiritual head of the Kagyu Order, has made very similar statements.** Both great leaders of Tibetan Buddhism promote vegetarianism, because the alternative involves doing harm to other beings. This represents a major development within contemporary Tibetan Buddhism, which signals an unequivocal return to the Buddha's original teachings.

There is no question about it, as Yoga practitioners we must take the traditional virtue of nonharming (*ahimsa*) very seriously. That means, we must inspect our dietary habits and, if necessary, revise them. All life forms deserve our respect.

---

*See http://www.shabkar.org/teachers/tibetanbuddhism/dalai_lama.htm.

**For the statement by Orgyen Trinle Dorje, the Karmapa XVII, see www.shabkar.org/teachers/tibetanbuddhism/orgyen_trinle_dorje.htm.

The German physician and theologian Albert Schweitzer, who won the Peace Nobel Prize for his humanitarian work in Africa, coined the phrase "reverence for life." We can think of no better expression to describe Yoga's fundamental attitude toward all beings. All life is sacred. Some people have a problem with the word "sacred," but "reverence," "respect," or "care" will serve just as well.

Mahayana Buddhism proffers the wonderful notion that all sentient beings—which, we propose, is really all living beings—are our Mothers. What this means is that we ought to treat all beings as we would a loving mother, namely with care and regard. To quote ecosophist and environmental activist Arne Naess again:

> I have injured thousands of individuals of the tiny arctic plant, *Salix herbacea,* during a ten-year period of living in the high mountains of Norway, and I shall feel forced to continue stepping on them as long as I live there. But I have never felt the need to justify such behavior by thinking that they have less of a right to live and blossom (or that they have less intrinsic value as living beings) than other living beings, including myself. . . . The broad stream of nature poetry, over thousands of years, is perhaps the best source of confirmation of the widespread intuitive appreciation of the *same* right of all beings to live and blossom.*

---

*Arne Naess, "Equality, Sameness, and Rights," in *Deep Ecology for the 21st Century*, ed. by George Sessions (Boston, Mass.: Shambhala Publications, 1995), pp. 223–224.

3 RIDING THE COSMIC WATERS

The Amniotic Fluid of the Universe

Hydrogen (H) is the lightest and also the most abundant element in the universe. It has a single electron. Most commonly it appears in a diatomic state (having two atoms), as H_2. Astrophysicists tell us that 75 percent of the mass of all matter in the cosmos is composed of hydrogen gas, which is odorless, tasteless, and colorless but highly flammable (as became evident when in 1937 the proud airship *Hindenburg* disintegrated rapidly before the eyes of thousands of spectators). Hydrogen is scattered throughout interstellar space at an extremely low density. In stars, like our own Sun, it exists in a plasma state and is so highly compressed that it converts to helium, which is the nuclear process that causes stars to shine so brightly.

On Earth, hydrogen is the third most abundant element, after oxygen and silicon. It exists only in various chemical combinations with other elements, notably hydrocarbons and oxygen. In combination with oxygen, hydrogen forms water (H_2O). Curiously, the surface of the Earth is covered by about 71 percent water, which also happens to be the approximate percentage of water present in the adult human body.

Remember the traditional dictum "As within, so without"?

The elements that exist all around us in Nature predictably also make up the human body. Here is the physical composition of our human microcosm:

1. oxygen (65%)
2. carbon (18%)
3. hydrogen (10%)
4. nitrogen (3%)
5. calcium (1.5%)
6. phosphorus (1.0%)
7. potassium, sulfur, sodium, magnesium, copper, zinc, selenium, molybdenum, fluorine, chlorine, iodine, manganese, cobalt, iron, lithium, strontium, aluminum, silicon, lead, vanadium, arsenic, bromine (collectively 1.5%)

From the perspective of the elements, we are mostly oxygen, but, as mentioned above, oxygen combines with hydrogen to form water and so, chemically speaking, we amount to a miniature ocean. This is fitting enough, because biologists inform us that life originated in the oceans. Without drinking water, physicians are convinced, we could not hope to live much beyond three weeks. Thus, water is indeed the stuff of life.

The superlative importance of water is captured in many of the world's creation myths. In the *Rig-Veda*, for instance, the created cosmos is often said to have its origin in the cosmic waters. Thus, the well-known Hymn of Creation (10.129) mentions the unfathomable oceanic depth (*ambhas*)—the infinity of transcendental "space"—giving birth to the world.

For the thinkers of the Vedic civilization, water had several levels of meaning. At the symbolic-metaphysical level, water signified the imponderable Reality prior to the emergence of space-time. At the psychological level, it denoted the fertile ground of the deeper mind (or unconscious). At the concrete level, water was seen as the great agent of purification.

In the last-mentioned capacity, water not only cleanses the skin of dirt but also purifies the body internally. "In the waters is ambrosia; in the waters is healing balm," notes one verse of the *Rig-Veda* (1.23.19). The purifying power of water extends to the emotional and spiritual dimensions of our being. We can readily appreciate this when, after a stressful day, we take a shower and wash away the accumulated external grime and internal "impurities," or tensions. We can also recognize the purifying, rejuvenating, and healing aspect of water when, on a hot summer's day, we drink a glass of cool fresh water to quench our thirst.

Given Yoga's orientation toward self-purification, it is not surprising that it should have developed an elaborate water symbolism and also carefully utilized water in its repertoire of practices. The texts of Hatha-Yoga, for example, recommend cleansing baths in a flowing river, sprinkling oneself with water, offering libations. All such practices are to be conducted in a ritualistic (mindful) fashion, paying due attention to the psychological component of water's cleansing power.

The *Gheranda-Samhita* (1.44), a classical Hatha-Yoga manual dated to c. 1650 A.D., includes water enemas and also nasal douches in its suggested routine. This text (4.22), moreover, furnishes the dietary stipulation that the Yoga practitioner should fill the stomach half full with food and one fourth with water, leaving one fourth empty. Sundaradeva's *Hatha-Tattva-Kaumudi* (4.22), dated c. 1750 A.D., specifically recommends the drinking of rain water.

All these traditional practices assume that the water used

is clean and fresh. Sadly, today, we can no longer safely take baths in rivers or drink rain water. Our rivers have become often heavily polluted, and in many areas rain has become so acidic that it is destructive of rocks and buildings, including time-honored antiquities like the pyramids of Egypt and Mexico.

Even the costly bottled water that so many Yoga practitioners carry around with them has been shown to be little better, or even worse than, the treated water flowing from our household taps. Apart from anything else, plastic bottles have a high environmental price tag. The bottling of water is now a $40 billion industry worldwide! Plastic bottles also are a serious health hazard, because they leach the toxic metal antimony into the water, which can cause nausea, dizziness, and depression and in larger doses can even be fatal. Another group of dangerous chemicals found in plastic-bottled water are phthalates, which play havoc with the endocrine system. In his travels as an environmental spokesman, David Suzuki insists on being given tap water rather than bottled water. He rightly points out that if we cannot trust our tap water, we ought to be "raising hell"* with municipal authorities.

As we will show next, water is no longer the water of life. In some instances, it has become the exact opposite: a bringer of ill health and death. Considering that physicians recommend that we drink anywhere from five to ten 8-ounce glasses of water a day, this represents a serious quandary. We begin our overview with the ocean, which evolutionarily has been the cradle of life on Earth. It also gives our planet its characteristic blue color when viewed from outer space—a telltale sign of possible life for any carbon-based extraterrestrials who might be passing through our solar system. On their approach to the blue marble hanging in space, however, they would also notice

*David Suzuki in an interview for CBC news, dated February 1, 2007.

the large brown haze of heavy pollution over portions of the land masses. This would suggest to them that the life forms on our planet are, after all, not entirely intelligent.

The Plight of the Oceans

The ocean, which forms a single continuous body of salt water but is generally divided into five regions (called "oceans"), covers an area of 361 million square kilometers (139 million square miles), or 70 percent of the Earth's surface. Its estimated volume is c. or 1,340 million cubic kilometers (319 million cubic miles). What an incredible vastness! And yet, unbelievably, we have managed to thoroughly pollute the Pacific, Atlantic, Indian, Southern, and Arctic oceans along with the various seas associated with them, such as the Mediterranean Sea and the Black Sea.

I (Georg) remember how dispirited I felt when, in my early twenties, I was reading Thor Heyerdahl's book *Ra* and came across his comment that he had watched garbage floating by his papyrus raft in the middle of the Atlantic. I could not fathom how such a vast expanse could show any sign of human-caused pollution. Since his daring *Kon-Tiki* expedition, which took him 4,300 miles across the Pacific in a balsa raft in 1947, garbage (mostly plastic) drifting in the ocean swells had increased exponentially. In the years since the *Ra* expedition in 1970, the situation has become even worse.

In 1997, Charles Moore, who has had a life-long love affair with the ocean, traveled on his 50-foot catamaran to a desolate area known as the North Pacific Subtropical Gyre. This is a stretch of ocean comprising about 26 million square kilometers

(10 millon square miles) where the waters rotate slowly in clockwise fashion. It is avoided by sailors for its lack of wind.

Moore steered straight for this vortex and discovered an ugly truth: The entire area had become a garbage dump for millions of floating plastic articles. Subsequent research revealed that the plastics were breaking down into minute fragments, which then become entwined with plankton, the food of numerous sea creatures. Now known as the "Garbage Patch," this nightmarish place is a veritable cocktail of toxicity. (We will say more about the lethal heritage of plastics in Chapter 4.)

In the meantime, investigations have identified several other similar gyres in the ocean where plastics mill around until they become a soup of minute death-dealing particles, destined to exist for thousands of years and progressively killing of sea life. What is particularly disgraceful is that most of the plastic items floating in the ocean stem not from ships but from the land.

In recognition of the many human-made problems connected with the ocean, the United Nations declared 1998 as the International Year of the Ocean to bring greater awareness to people worldwide. As far as we are concerned, this should be an annual recognition, because the ocean is the matrix of life. When the ocean dies, we all die. What, then, are those problems?

As you might have guessed, a major contributor to oceanic pollution is shipping, which makes up c. 90 percent of all trade by volume between the 25 member states of the European Community and c. 80 percent of all goods shipped in and out of the United States.* Shipping accounts for about 12

---

*See the report *Air Pollution and Greenhouse Gas Emissions from Ocean-going Ships: Impacts, Mitigation Options and Opportunities for Managing Growth* published by The International Council on Green Transportation, March 2007. Available online in PDF format.

percent of the total marine pollution due to human activities. It involves nasty polluting practices, such as the discharge of ballast water; disposal of marine debris and waste materials as well as sewage; oil spills and spills of hazardous chemicals, including—believe it or not—radioactive materials.

As if this catalog of environmental flaws were not enough, shipping also contributes to air and noise pollution. As for the former, the role of shipping in air pollution has been properly evaluated only in recent years. Incredibly, the 90,000 or so ships plying the oceans release more sulfur dioxide than all cars, trucks, and buses on Earth combined. In addition, these ships are responsible for a sixth of nitrogen oxide pollution worldwide. Also, carbon dioxide emissions from the international shipping sector exceed the total greenhouse gas emissions agreed on by most of the industrialized nations of the Kyoto Protocol (1997). So much for maritime commercial shipping.

Cruise ships are in a class all their own as major polluters. Thanks to Ross Klein, a professor of social work who has taken over thirty cruises, everyone can have a much clearer picture of the ugly side of this growing industry.* We feel obliged to single it out, because it has become rather popular to go on "Yoga cruises" with the stars of contemporary Yoga. We imagine that few, if any, of the Yoga teachers and students involved in this latest fashion know just how destructive cruises are. Here are some facts to think about.

---

*See Ross A. Klein, *Cruise Ship Blues: The Underside of the Cruise Industry* (Gabriola Island, B. C.: New Society Publishers, 2002) and *Cruise Ship Squeeze: The New Pirates of the Seven Seas* (Gabriola Island, B.C.: New Society Publishers, 2005). We would like to note here the incredible work done by New Society Publishers in making critical exposes in diverse sociocultural areas widely available.

Cruise ships regularly dump oil and hazardous chemicals, as well as waste water and sewage into the ocean and, if caught, the cruise line pays the comparatively modest fines rather than change its policy. According to the U.S. Environmental Protection Agency, each passenger generates c. 378 liters (100 gallons) of waste water, including 10 gallons of sewage. Multiply this by several thousand passengers per ship. Passengers also generate burnable waste, food waste, and waste in glass and tin totalling c. 3.5 kilograms per day, which all too often end up at the bottom of the ocean. This boils down to tens of thousands of tons every year! There is simply no excuse for such wanton littering and contamination of the ocean.

Switching from an environmental to a social perspective, we need to know that employees working on cruise ships typically put in a 16-hour day, seven days a week, at below minimum wage while being housed in substandard accommodation. In other words, they slave to cater to passengers' penchant for self-indulgence and to the corporate owners' greed. In 2004, the North American cruise industry made a net profit of over $2.5 billion!

Ever since the late 1950s when transoceanic travel became popular, the cruise industry has been construction ever bigger ships. Today, ships of 100,000 tons and upward can carry 3,000 and more passengers. The cruise lines are constantly competing for the most tonnage and the biggest passenger capacity. Carnival Cruise Lines, the largest in the world, is currently constructing a 170,000-ton ship with a capacity for 5,000 passengers and 1,500 crew. This size ship represents a small town floating in the ocean at an enormous (largely hidden) expense to the maritime environment.

Unless governments intervene, virtually none of the industry's huge profits will go toward eliminating pollution or clearing up the formidable mess left in the wake of cruise ships. As with other industries, the pressure to conform to hard-won existing environmental standards and to put into effect better standards must come from the public. You and us!

In his books, Ross Klein offers a slew of other alarming data, including inadequate hygiene, healthcare, and security on board of cruise ships. Food poisoning, theft, sexual harrassment, sexual exploitation of female employees, and rape are as common on ships as they are on land in communities of equal size, but without comparable precautions and protection.

The most significant source of oceanic contamination are the countless polluted rivers around the globe that pour into the ocean. We will turn to them next.

Rivers and Lakes of Death

Our planet is estimated to sport upward of 100,000 rivers. The Amazon, the largest (though not the longest) river on Earth, alone has more than 1,000 tributaries many of which, in turn, have their own tributaries. The Mississippi River, which drains c. 40 percent of the continental United States, has the unenviable distinction of having been the most polluted waterway on the planet until recently, spilling annually some 1.5 million metric tons of nitrogen pollutants into the Gulf of Mexico and creating a coastal dead zone. Every year, U.S. waterways are burdened with having to absorb some 1.2 trillion gallons of untreated sewage, agricultural, and industrial waste.

In the meantime, as a result of China's rapid industrialization, the Huaihe River has overtaken the Mississippi in its pollution index. It is severely polluted with petroleum, ammonia, nitrogen, permanganates, volatile phenols, mercury, and other contaminants that asphyxiate aquatic life. The other great waterways, such as the systems of the Yangtze River, Yellow River, Pearl River, and Songhuajiang River, are not far behind.

China, which also has seven out of the ten most polluted cities in the world, is facing enormous environmental degradation, which should be a warning to all other developing nations. Potable water has become an urgent priority, as 90 percent of the country's urban water is polluted, and according to Xinhua News Agency, a governmental official is on record as stating that annually 410,000 people are direct victims of pollution—which is likely to be a gross understatement. Our reason for saying so is that, according to the Chinese Ministry of Water Resources, over 300 million residents in rural areas do not have access to safe drinking water, and about 190 million people drink contaminated water that is known to be harmful to their health.

Access to water in general has become a problem in the People's Republic of China, which is one of the reasons why that nation has constructed no fewer than 19,000 large dams, as compared to the United States' 5,500 and India's 4,300. Dams, however, are greatly devastating to the environment, partly because of the displacement of humans and nonhuman beings alike (including plant life) and the loss of land and livelihood, and partly because of their role in spreading waterborne diseases. Another reason for building large dams is of course people's insatiable appetite for electricity on the part of the industrialized or industrializing world.

The situation is lamentably similar in India, which is next to China the only other nation with a population of over one billion people. Fourteen percent of the total length of the Ganges River, held sacred by millions of Hindus, with its more than 3,700 miles (or 6,000 kilometers) has been classified as severely polluted. Nineteen percent of all others rivers are considered moderately polluted. Domestic sewage and industrial pollutants are the leading contaminants of rivers, where they cause oxygen starvation. They also often seep into precious groundwater, causing a serious shortage of potable water. Astonishingly, the Central Pollution Control Board of India is considering an ambitious plan that would link 37 major rivers in what could prove a lethal network of crosspollution.

China and India are geographically far away from North America, which most readers of this book are likely to call their home. Their problems may seem rather remote and irrelevant to us. This, however, is an entirely wrong perspective. We can no longer afford to be ethnocentric. **The pollution of other countries is also ours, and our pollution is theirs. We inhabit the same biosphere, which is in the process of collapsing.** We all, especially those of us living in industrialized or developing nations, are co-responsible for the present state of the world and also for the necessary cleanup. The idea of humanity being one family is not just a pious slogan but a reality that now must inform all our individual, social, economic, and political actions.

Thus, practically speaking, when we buy consumer goods manufactured in China or India or anywhere else in the world where environmental standards are low, we inevitably contribute to global pollution. In fact, we do so every time we purchase anything that involves long-distance transport on land or water or by air.*

River water comes primarily from stored water (ground-

*See our discussion on the 100-mile diet in Chapter 5.

water, lakes, or glaciers) or from runoff due to precipitation. All these sources have been found to contain human-caused contaminants. These become concentrated as rivers flow downhill toward their final destination, be it the ocean or one of the numerous lakes. At this point, all of Earth's countless rivers are more or less polluted; mostly more rather than less. Many big rivers terminate in the ocean, where they contribute to further marine contamination.

The earliest human civilizations all arose on the banks of major rivers—the Indus and Sarasvati in India, the Nile in Egypt, Tigris and Euphrates in Mesopotamia, and the Yangtze in China. They fertilized the soil, provided a source of food and potable water, and served as avenues of transport. Today rivers are almost universally showing signs of environmental corruption, and their once life-giving role is swiftly reversing into its opposite.

Streams, rivers, and lakes are fast becoming lethal to aquatic and human life. It is hard to believe that 40 percent of the rivers and nearly half the lakes in the United States, for instance, should be considered unfit for fishing and swimming.

Even back in the late 1960s, when hiking in the highlands of Scotland, I (Georg) still felt fine about sipping the cool and delicious water bubbling forth from a spring high up on a large moss-covered boulder. When I (Brenda) used to go camping in the early 1970s, I freely drank the water from lakes and rivers in Northern Saskatchewan, as did other children. None of us ever experienced any ill effects from this. Nowadays, no reasonable person would want to take a sip from a stream or river without knowing what is going on upstream. As the next section shows, even then there is a risk.

Acidity From Above, Radioactivity From Below

Precipitation in the form of rain, snow, or dew, as already mentioned, is one of the ways in which rivers replenish themselves. Scientists have known about the phenomenon of acid rain, that is, rain with a low pH value, since the mid-nineteenth century.* It was one of the unforeseen side effects of the Industrial Revolution and one that has proven difficult to control, never mind eliminate.

Just as we use vinegar to curdle milk or clean brass, acid rain has a similar effect on the environment. It leaches away essential nutrients and minerals and kills off fish, birds, and insects, and it has been linked to asthma and other symptoms of ill health in humans. Acid rain, moreover, has a deleterious effect on buildings, including treasured historical monuments.

The main culprits in acid rain are sulfur and nitrogen compounds, such as those released naturally in volcanic eruptions and those created artificially and released from the smokestacks of factories and power plants and from the exhaust pipes of hundreds of millions of automobiles around the globe.

Rain, once thought a blessing, has become a questionable ally and, at times and in many parts of the world, a veritable ene-my. For 10,000 years, the carbon dioxide content of the atmosphere remained stably at approximately 270 parts per million. Ever since the Industrial Revolution, we have increasingly and liber-ally burned coal, oil, and gas to satisfy our ever-growing need for energy. As a result, we have polluted air, water, and soil, and set in motion a chain reaction summarized under the label

*The pH value of a liquid indicates its alkalinity or acidity on a scale from 0-14. Whereas pH7 is neutral, vinegar, for example, which has a pH value of 2.4, suggests high acidity. Acid rain is generally defined as rain with a pH value of less than 5.6.

"global warming." Acid rain is one of the disquieting consequences of a long line of misjudgments and misapplications, which are now among the factors contributing to the collapse of our biosphere.

If acid rain is gnawing on us from above, radioactivity of human origin is clawing at us from below. Earlier in this chapter, we referred to radioactive waste as being among the pollutants of the ocean. With the Cold War between the United States and Russia having receded into history, most people seem to have forgotten about the dangers of nuclear war and radioactive fallout. It is a terrible fact, however, that the above-ground and underwater nuclear tests so irresponsibly conducted by the United States between the late 1940s and the 1960s have left their indelible mark on the environment.

Out of ignorance and, we would argue, out of callous indifference, nuclear technicians and their administrators chose the ocean for their dumping ground. Between 1946 and 1970, the United States dumped some 47,800 large containers of radioactive waste in the ocean west of San Francisco—a desolate area of c. 1400 square kilometers (540 square miles) known as the Farallon Island Radioactive Waste Dump.

Other nuclear nations have been guilty of the same irresponsible disposal of radioactive waste. Thus, the former Soviet Union is known to have dumped radioactive waste in the Sea of Japan between 1950 and 1993. Just one month after Greenpeace had exposed this unconscionable practice, Russia defiantly dumped another 900 tons of low-level radioactive liquid in the Sea of Japan. Responding to Japan's outrage, the United States and others nations finally took diplomatic steps to

---

*See, e.g., T. J. Freeman, ed. *Advances in Underwater Technology, Ocean Science and Offshore Engineering*. Vol. 18: *Disposal of Radioactive Waste in Seabed Sediments* (Boston, Mass.: Graham & Trotman, 1989).

stop this reckless pollution.

Recently, it also has come to the public's knowledge that for several decades the former Soviet Union dumped much of the spent fuel from its nuclear-powered vessels into the Barents and Kara Seas, 7,000 tons to be precise. Both Seas are situated north of the Russian town of Murmansk, where 100 tons of nuclear waste are very inadequately stored, having the potential for a large-scale environmental disaster.

Nobody knows just how many oceanic dumping grounds there are. In the meantime, scientists are busy pondering how to get rid of the ugly evidence of their Faustian pursuits. Would you be surprised to learn that, once again, they favor the ocean as the destination for nuclear waste? For lack of a better alternative, some of the nuclear nations are indeed considering the option of burying nuclear waste deep in the sediment of a supposedly stable sea floor where, they hope, it will be carried deeper.

Some governments, notably the United States, Canada, and Great Britain, are poised to bury high-level nuclear waste underground in deep caverns or shafts. The problem is that no known material is immune to corrosion or disintegration for the span of 10,000 years. No one can possibly foresee what tectonic movements might occur even in apparently stable geological structures. Until an acceptable solution has been found, highly toxic materials are stored in facilities where they are vulnerable to leakage and terrorist attacks. We are bequeathing to future generations a most hazardous and unpredictable legacy. We will say more about radioactivity in Chapter 4 when discussing the pollution of land and soil.

Only by correcting our faulty attitudes toward the ocean, lakes, and rivers and their multitudes of beings can we hope to secure the water we need for our own survival. As Rod Fujita, senior scientist with Environmental Defense in Oakland, California, and one of the United States' foremost leaders in marine conservation, wrote:

> We need to develop "ecolacy" to complement the essential modern survival skills of literacy and numeracy. Ecolacy is . . . the prudent practice of asking "and what then?" of any technological innovation or economic activity.*

When the ocean, lakes, and rivers die, so do we. When they thrive, we will thrive as well. And only when we thrive will we be able to discover the hidden dimension of the water element, which is the "cosmic water," our universal birthing ground, as we can experience it in the state of ecstatic communion with all of life.

Deadly Water of Life

Next, we must turn our attention to the growing shortage of potable water on Earth. This has become a global problem. It is already a serious crisis for many Third World countries and nations with large and rapidly expanding populations, such as China and India. Before very long, as the world population continues to swell, the shortage of drinking water is expected to assume critical mass for all nations. By 2025, forecasters predict, two thirds of the world's population will not have access to safe drinking water. It is an undeniable fact that the

*Rod Fujita, *Heal the Ocean: Solutions For Saving Our Seas* (Gabriola Island, B.C.: New Society Publishers, 2003), p. 4.

Earth with its over 6 billion people has reached and even gone beyond its "carrying capacity."

Fresh water constitutes a mere 3 percent of all water on Earth. To get an idea of how minute a quantity this is, fill an aquarium with 20 liters of water. Of this, 19.5 liters (or 97.5 percent) represent the salt water in the ocean. Three quarters of the remaining 0.5 liters is unusable water found in the soil or water frozen in glaciers and the ice caps. One eighth represents unreachable or unusable groundwater. The remaining eighth is drinkable water, which would be about 5 drops of a 20-liter tank, or 0.007 percent. Little wonder that potable water has been called "blue gold" or "liquid gold."

According to the World Health Organization, over 1 billion people already are lacking access to potable water and are reduced to drinking unsanitary and unsafe water. Since these are typically poor people whose health is already compromised, this situation represents an added health hazard. Various water-borne infectious diseases are the principal cause of morbidity and mortality in Third World and developing countries. *Every day*, 5,000 or so children are dying from contaminated water—as many children under the age of 15 as are killed in gun-related violence in the United States every year; or as many children from Russia and Eastern Europe as are adopted by Americans annually.

Many countries are digging deeper and deeper wells to tap into groundwater resources and in the process manage to overexploit existing aquifers that might have taken thousands of years to form. As a consequence, groundwater levels keep dropping precipitously. For example, in only 35 years, Northern China's groundwater overdraft has reduced water levels by 50 meters. Some regions of India are facing the same problem.

Not a few analysts think that this kind of global water stress could lead to war between nations. Conflicts over water appear to be as old as human civilization itself. Countries suffering from lack of adequate water resources or rainfall, such as those in the Middle East, have known water-related conflict for millennia. Significantly, when the former Egyptian president Anwar Sadat signed the 1979 peace treaty with Israel, he remarked that his country would never go to war again—except over its precious water resources. Should this comment not have been taken into account when the Swedes awarded Sadat the Peace Nobel prize in 1978?

When Turkey built a huge dam, which now regulates the flow of water in the Tigris River, both Iraq and Syria were understandably alarmed. Their agriculture depends on the alluvial plain between the Trigris and Euphrates. Who owns a river? Or a lake? Or the ocean?

We don't need to look as far afield as the Middle East to identify potential flashpoints over access to water. Intense conflict has existed for some time already between Canada, which is blessed with 20 percent of all fresh water on the planet (mostly in the form of the Great Lakes), and its southern neighbor, the United States.

Ever since its inception, the United States has coveted Canada with its colossal natural resources. It even invaded Canada in 1812 but fortunately failed in its mission to conquer this peace-loving country. In 1974, declassified top-secret documents shockingly revealed that in the era between the two world wars, Washington had drawn up plans for yet another military invasion. This thought seems to rear its head periodically with some American politicians, and the Great Lakes are never far from their mind.

Within the United States itself, water wars have been fought, figuratively but also literally, for years. Thus, in California, the populous city of Los Angeles, which is located in an arid part of the country, has been fighting the residents of Owens Valley over water since 1898. At one time, the Los Angeles authorities extracted so much water from Owens Lake that the local farmers took up arms. Not surprisingly, the metropolitans won, and agriculture disappeared from Owens Valley. After a second aqueduct was constructed in 1970, which drew primarily from groundwater, the local flora began to wither as well.

This particular conflict epitomizes the age-old rift between urban dwellers and farmers. What the former often don't seem to appreciate is that without water for farming, they would not have any food to eat. At the same time, however, present-day farming practices, especially cattle farming, are wasteful of water, just as the urban lifestyle tends to be extravagant in its use of natural resources.

Elsewhere in the United States and also in other parts of the world, tension over access to water, especially drinking water, has intensified over the past couple of decades. According to the U.N. Environment Agency, 3 billion people will be without adequate water—both in quantity and quality—within the next 50 years. From the available evidence, we think that this is an overly optimistic forecast. Already over 1 billion people are without adequate water, and it won't take very much to add another couple of billion people to this unfortunate group.

Smelling huge profits and often supported by national and local bureaucrats that keep their eyes fixed on economic growth or on padding their own pockets, multinational corporations have mobilized to buy up local water rights from state and municipal authorities. They are not even trying to conceal the fact that they are seeking to monopolize fresh water. All this flies under the legal banner of "privatization" of water.

This business strategy has proven particularly noxious overseas, where local governments provide foreign corporations with the necessary military backup to silence the protesting population. Thus, the San Francisco-based multinational giant Bechtel, which has an annual revenue of around $20 billion, had taken over the water supply of Cochabamba, Bolivia—a forty-year deal. The price-hiking that followed caused a revolution among the locals, who depend on cheap water, and in the end Bechtel had to withdraw. In the meantime, this corporation, which specializes in mammoth construction projects around the world, even managed to alienate San Francisco's Board of Supervisors for overbilling the city, where the company is headquartered.

Few water wars end relatively favorably for the local people. Some foreign corporations have demonstrated a no-holds-barred approach to get access to water (and other natural) resources and to remain in control of them. Protesters have been intimidated, wounded, and even killed. All they were asking for was to have access to affordable safe water or to not have their traditional lifestyle undermined or their land destroyed.

After a run of bad experiences with multinational corporations, more governments in developing nations are responding to public pressure by voting against the privatization of water. Corporations treat water as a commodity to be converted into easy profit. For ordinary individuals, water is simply a means to quench their thirst, cook, bathe, or wash their clothes. As nations and individuals, we evidently must learn not only to become wise consumers but also eager collaborators with our neighbors.

Looking at consumption patterns, developed nations are flagrant violators of the age-old principle of moderation. The United States dubiously leads the world in per-capita water consumption at the rate of c. 2,000 cubic meters (or 70,000 cubic feet) per year. The Canadians are second, with a rate of c. 1,600 cubic meters (or 56,000 cubic feet), as compared to c. 708 (or 25,000) of the average German and 254 (or 9,000) of the average Dane. Europeans seem to be more conscientious about water usage but North Americans are veritable squanderers.

About one thing there is no doubt: Water consumption is outpacing available water resources. At present, the United States relies mostly on the huge Ogallala aquifer, which stretches over a distance of 1300 kilometers (800 miles) from Texas to South Dakota and comprises c. 451,000 square kilometers (174,000 square miles), which translates into 4 trillion tons of water. Every year, some 12 billion cubic meter (420 billion cubic feet) are pumped out of this vast underground reservoir, irrigating roughly one fifth of the country's agricultural land. According to some estimates, this gargantuan source of fresh water, which accumulated some 2 to 6 million years ago, may essentially run dry as early as 2030 or 2040 if the present rate of usage continues—an all-out disaster in the making. But even 100 years would be far too soon.

Given Washington's historical record of imperialist politics and long-standing designs on taking over Canada, the Canadians are naturally worried that their neighbor to the south won't just purchase water to satisfy its ever-growing thirst but hijack the Great Lakes or aggressively pump water from the U.S. portion of the lakes. Since the Great Lakes are interconnected, the latter option would automatically lead to a drop in the water levels of the Canadian-owned parts of the lakes as well. This is particularly worrisome in the case of Lake Ontario, where overpumping could conceivably dry up the St. Lawrence River running through Quebec. The river is a source of food and livelihood for millions of Canadians, and a major transport artery.

Take the massive Colorado River, which has its headwaters in the Rocky Mountain National Park of Colorado and meanders all the way to Mexico, some 2330 kilometers (1450 miles) to the south. Americans draw water so liberally from this river that by the time it crosses the border, the Mexicans are left with about one tenth of its full size. Most of the time the river's fresh water no longer reaches the Gulf of California but runs dry in the Sonora desert. This has proven catastrophic for the once rich flora and fauna of the Colorado River Delta.

Such critical issues are bound to rear their head more and more frequently not only on the American continent but throughout the world. Again, unless sanity prevails, they will predictably lead to conflict and possibly war.

Finally, we must ask what North Americans, who have such lavish fresh water resources available to them, do with the water that is pumped into their homes. You would imagine that it is delivered to everyone in the purest state possible both for enjoyment and health. Not so. In many states, the water first undergoes a fairly complex purification process that is intended to remove unwanted micro-organisms and chemicals to give people the best possible water quality. In practice, however, this process—which is controlled by municipalities—is varyingly successful.

In many North American municipalities, the water flowing from people's taps is full of iron and aluminum (two metals added in the coagulation phase of the purification process, which is supposed to eliminate organic matter), and not least chlorine (for disinfection). Now, stained sinks and toilets caused by iron in the water are unpleasant to look at, but chlorination is a distinct health hazard. If you have ever dipped into a public swimming pool, you will know that chlorine stains your swimsuit and your hair green, hardly an encouraging sign. In fact, chlorinated water has been associated with cancer (especially of the bladder) and heart disease.

If you happen to live in an older home, your water is likely to run through lead pipes and thus will carry a fair dose of lead, which is a toxic substance that can damage the nervous system, reproductive organs, kidneys, and brain and cause anemia. The problem is serious enough for the World Health Organization to pay attention to it.

Another potential health hazard is fluoridated water, which nowadays is delivered to nearly 70 percent of Americans and 40 percent of Canadians. In Europe, only Ireland has a mandatory fluoridation program. Despite the warnings of natural health practitioners, many municipal authorities still fluoridate the water, for no other reason than to fight tooth decay. While fluoride is found in all water, the practice of fluoridation, which dates back to the mid-1940s, increases the natural level to 0.7–1.2 parts per million. Of course, when the fluoridation equipment is faulty, as has happened, the water becomes poisonous.

But some physicians have made a case that all fluoridated water is toxic to the human body, particularly the nervous system. Fluoridated water has also been associated with bone cancer in children and uterine cancer, and so forth. In 1997, more than 1500 professionals at the United States' Environmental Protection Agency made a public statement against the addition of fluoride in California's water supply, because this would make the drinking water unsafe.* This expert vote and local citizens' protests notwithstanding, fluoridation continues in California as elsewhere. Moreover, the toxin is freely available in toothpaste and even in so-called "health" products. We rest our case.

---

*For an excellent discussion and exposé, see Christopher Bryson, *The Fluoride Deception* (New York: Seven Stories Press, 2004).

Healing Waters

This chapter has been piling bad news upon bad news. Is there, then, no good news at all? There is *some,* and a few years ago, David Suzuki made a point of publishing a work entitled *Good News For a Change.** One of the most difficult aspects of working on this book has been the ample daily fare of heart-wrenching bad news coming across our desks. At times, we have had to take a break from sifting through the news as it was coming in, not merely to assimilate and appraise what we were reading but also to allow our feelings of sadness or anger to settle and recompose ourselves.

Compared to all the bad news, the good news tends to arrive in a mere trickle. Thus far, it has always managed to restore our faith that, if enough people were to truly comprehend the ongoing environmental devastation and become active in taking corrective measures, the situation could be turned around. We are under no illusion, however, that even if everyone on Earth were to adopt a sustainable lifestyle today, we could collectively stop the downward slide *immediately.*

A moving truck or train takes time to slow down even when the air breaks are applied. **We have to reckon that the present-day environmental degradation has already gathered sufficient momentum for any positive course of action to take decades to correct what is happening, which is why we need to act NOW. Within Nature's time frame, which is calculated in millions of years, a few decades or centuries are a rather short period. In human terms, this gradual turnabout will greatly tax the average person's patience and commitment**

*See David Suzuki and Holly Dressel, *Good News For a Change: How Everyday People Are Helping the Planet* (Vancouver: Greystone Books, 2003).

to sustainability.

Our consumer-oriented society, unfortunately, has made us prone to expect instant gratification of our desires. Environmental change, which involves countless intricate and interconnected factors, is a long-term affair. One of our concerns is that people may not have the staying power when their green activism fails to yield desirable results promptly enough, especially when we start to seriously experience the natural forces unleashed by our careless actions over the past 100 years and more, notably the ill effects of global warming.

As psychologists confirm, most people manifest considerable emotional immaturity, which is definitely not in favor of the kind of responsibility and long-range perspective that are called for today. Therefore, in our opinion, unless governments introduce parental measures of control not in the interest of the corporate world or consumerism but in the interest of all sentient beings, we are unlikely to see large-scale concerted green action on the part of the majority of people in the near future.

We can, however, expect to see more and more individuals worrying about the environment and the future and as a consequence adjusting their lifestyle more or less drastically. We hope that this contingent of humanity will come to include every single Yoga practitioner.

Among positive efforts to heal the environment, we want to mention all those many volunteers in different parts of the world who for years have collaborated to restore wetlands and estuaries, clean up lakes and rivers from pollution or beaches after oil spills, protest the building of dams, confront whalers, lobby local and national politicians for save drinking water, or campaign against the privatization of water, and so on. Without their sustained endeavors and personal courage, we would all be worse off.

Action is needed now, and the only responsible thing you, as a concerned individual, can do is to reduce your own [greenhouse gas] emissions as far and as quickly as possible.*

—Tim Flannery

With six billion people on the planet, the risks are everywhere apparent. A disruption in monsoon patterns, a shift in ocean currents, a major drought—any one of these could easily produce streams of refugees numbering in the millions. As the effects of global warming become more and more difficult to ignore, will we react by finally fashioning a global response? Or will we retreat into ever narrower and more destructive forms of self-interest? It may seem impossible to imagine that a technologically advanced society could choose, in essence, to destroy itself, but that is what we are now in the process of doing.**

—Elizabeth Kolbert

We are a daring civilization. An adventurous civilization. But we are also a stupid civilization, afflicted with a death wish. We are destroying (not just using, but destroying) natural resources and habitats that are the very tissue from which society is made . . . We are destroying life at large.***

—Henryk Skolimowski

---

*Tim Flannery, *The Weather Makers: How We Are Changing the Climate And What It Means For Life On Earth* (New York: HarperCollins, 2005), p. 301.

**Elizabeth Kolbert, *Field Notes From a Catastrophe* (London: Bloomsbury Publishing, 2006), p. 187.

***Henryk Skolimowski, *Eco-Philosophy: Designing New Tactics For Living* (London: Marion Boyars Publishers, 1981), p. 106.

4 IMMORTAL LAND, IMMORTAL FOREST

Our Lifestyle Is Killing Everyone

Our fire-making Homo erectus ancestors, who migrated out of Africa some 1.2 million years ago, were hunters and gatherers. Like cattle, they moved around in search of food, not least meat. Their native intelligence, however, made them far more adaptive and inventive than other mammals. By 30,000 B.C., the Cro-Magnon branch of Homo sapiens, the successor lineage of Homo erectus, was honing its cultural skills in the climatic challenges of the concluding era of the last Ice Age.

As soon as the ice sheets and glaciers began to retreat and rain fell again in abundance some 10,000 years ago, human creativity burst forth. Fertile land, especially along river banks, invited humanity to adopt a sedentary lifestyle. We amazingly quickly invented agriculture, large-scale food storage, androcentric religion with a priestly elite, kingship, urban life, division of labor, bureaucracy, writing, jurisprudence, war—*ergo* civilization.

Over the next several millennia, the human population multiplied and spread out into every corner of our planet. The Earth's total land area is c. 149 million square kilometers (or 57.5

square miles), which amounts to 29.2 percent of the total surface of our planet. It includes large areas that are entirely or mostly uninhabitable, such as the Arctic, Antarctic, Greenland, high mountains, and deserts.

The Neolithic population is reckoned to have been about 3–5 million people; by 950 A.D. the population of the world had jumped to 250 million; by 1802 to 1 billion; by 1928 to 2 billion; by 1961 to 3 billion; by 1974 to 4 billion; by 1987 to 5 billion; by 1999 to 6 billion. Today (mid-2007) it is over 6.6 billion, with about 90 million people being added annually!

We have discussed the problem of overpopulation briefly in Chapter 3 and made the point that the present world population is simply unsustainable. In other words, our planet does not have the resources to feed our species in such large numbers on a more permanent basis. This is especially true because the extravagant lifestyle of consumers in the industrialized world, which accounts for about one sixth of humanity.

It is the consumers who, propelled by the "need for more" and cleverly manipulated by omnipresent commercial advertising, drive the economic engine of industrialized society. They also are largely responsible for the environmental crisis. To be sure, industrial nations have their own share of poverty (which happens to be shamefully high in the United States).*

The industrializing nations, notably India and China, are growing their own consumer class, which, eyeing the material plenty of Western countries, aspire to the same lifestyle. With even fewer environmental standards in place than in the Euroamerican world, the burgeoning industries of those two expanding population giants, can be expected to exacerbate

*The poverty rate for the United States, arguably the wealthiest nation on Earth, is still a whopping 12.6% for the year 2005! Six million people joined the ranks of the poor between 2000 and 2005!

the environmental crisis sharply in the coming years. David Suzuki recently calculated that the Earth could deal with a carbon dioxide output of 1 ton per person. Our actual carbon footprint is very much bigger. In Canada it amounts to 17.5 tons, in the United States to c. 20 tons, and in the small Euro-pean nation of Luxemburg to c. 22 tons.

Our overconsuming and inconsiderate lifestyle is carbon intensive. And carbon dioxide is the chemical equivalent of a mass murderer who is stalking us by collective consent.

Since Cain Tilled the Soil

According to the Biblical book *Genesis*, Adam and Eve had two sons—Cain and Abel. While Abel was a herdsman, Cain took up the innovate occupation of farming. He killed his brother, because God had accepted his sibling's sacrifice but not his own. This fratricide proved symptomatic of the aggressiveness ("territorial imperative") spawned by a sedentary lifestyle involving numerous people crammed into a small space. For his evil deed, Cain was condemned by God to a life of wandering.

This curse caused Cain, in turn, to invent urban life and civilization. East of Eden in an area called Nod ("Shaking"), he founded the first town, which was named after his first-born son, Enoch. What is no myth is that the earliest cities were dependent on nearby agriculture. The literally ground-breaking discovery of the Neolithic—irrigation agriculture—turned arable land into prime value and power. The annual agricultural surpluses allowed a percentage of the population to stop farming and herding and instead specialize in various professions, such as potter, weaver, carpenter, toolmaker, baker, wine presser, artist, and not least administrator and priest.

To amass wealth, the cities strove to acquire agricultural surpluses and tradable commodities. The peasants cleared, tilled, seeded, irrigated the soil, and harvested the crops, while the city administrators or merchants stored any excess. As cities grew wealthier, they came under threat from envious neighboring communities and ambitious rulers. Raids and wars often undid the cultural, social, and economic advances made locally. Successful cities expanded beyond their original walls, which necessitated the clearing of more land, often leading to widespread deforestation and sometimes even local climate change.

Although the land surface of the Earth is truly vast, it is far from infinite. The poets' evocative phrase "immortal soil" is as misleading as the scientists and politicians catchphrase "progress." To feed over 6.6 billion mouths, especially with so many given to a taste for meat and animal products, places a huge strain on the land. The available arable land on Earth amounts to roughly 31 million square kilometers (12 million square miles). This translates into 0.005 square kilometers or 1.16 acres per person. In our time, an estimated 15 percent of soil is degraded. This includes c. 100,000 square kilometers (36,610 square miles), which are lost annually due to erosion by water (56 percent) and wind (28 percent). The precise amount of soil erosion resulting from tillage and urban development is unknown. To be sure, the major culprit of soil loss or degradation is deforestation, which we will address later in this chapter.

Modern agriculture, with its intensive technology, is essentially ruinous for the land. Overfarming can be as deleterious as deforestation. It bleeds nutrients from the soil and saturates it with pesticides instead. More and more, farmers have to resort to nitrogen fertilizer to restore a semblance of fertility to the land. The gain in dollar profits is offset by the environmental cost in producing, transporting, and using fertilizer. Thus, each ton of nitrogenous fertilizer requires a little over 948 cubic meters (33,500 cubic feet) of natural gas for its production. Although gas is easier on the environment than oil or coal, it is still a source of pollution.

Chemical fertilizers typically do not replace the trace minerals in the soil, which have been depleted through overfarming. Consequently the produce also lacks in trace minerals and therefore is less nutritious. Another side effect of chemical fertilizers is that they tend to strengthen certain types of pests, which, in turn, adversely affect the production and health of crops.

The frequent use of phosphate fertilizers, again, can lead to an accumulation of undesirable trace elements, such as arsenic, cadmium, and uranium, in the soil and thus also in the produce that we eat. Even small amounts of cadmium can prove harmful to one's health. At this point, much of our food is toxic, either because of the way it was grown or because of its packaging.

In the production of phosphate fertilizers, each ton of phosphorous acid generates 5 tons of phosphogypsum, which needs to be stored safely because of its radium content. Florida alone has amassed some 700 million tons of this toxic by-product! We think this is plain crazy.

Overfertilization, which is common in industrial-style farming, can lead to nitrogen or phosphorus runoff into waterways (and ultimately into the ocean), which then become hypoxic (or lacking in oxygen) killing fish and other life forms.

Nitrogen-based fertilizer is big business—over $20 billion around the world—which boosts corporate farming that is so destructive of the much more environment-friendly traditional family farming. Because of rising gas prices, some corporations have switched back to coal, which is gasified. Again, the hidden environmental costs are high.

Given all the problems with chemical fertilizers, it is amazing that not more farmers are choosing to go the route of organic fertilizers. Yet, outside North America, this is increasingly seen as a far better option. Thus, the International Federation of Organic Agriculture Movements founded in 1972, now has a membership of 670 organizations and institutions hailing from over 100 countries.

Even though Germany, for instance, is the home of many transnational chemical corporations, it has 8,000 organic farmers. Austria has more than 20,000 organic farmers, while Italy has 18,000 farms that are either fully organic or transitional. Great Britain is the largest trader in organic food. Other European countries are following suit, and several African countries (notably Uganda and Egypt) and South American nations (particularly Mexico) also show an increase in organic farming, which is very encouraging.

The organic food sector is the fastest growing sector of the food market and according to some estimates already amounts annually to about $100 billion worldwide. In Canada, the demand for organic food is outpacing production, even though there are some 15,000 farms dedicated to organic crops. In the United States, 4 million acres out of 938 million acres are certified organic (including organic animal husbandry). The worldwide figure for organic farming is around 75 million acres. This is encouraging but, again, much more needs to be done to boost organic farming in all agricultural countries of the world.

We cannot leave the topic of agriculture without also mentioning one of the most insidious and, in our view, sinister trends that must be stopped at all cost. This is the push toward genetically engineered crops, particularly nonreproducing seeds that force farmers to purchase seeds anew with every growing season.

Bioengineered crops are basically untested and usually are released to the public without due notification or proper education in the possible long-term consequences of such an approach, which was initially developed and supported by the U.S. Department of Agriculture.

As for nonreproducing seeds, these have rightly been dubbed "suicide seeds" or "terminator seeds," which are patented bioengineered products. These genetically modified seeds, which are hugely profitable, are good for a single generation of crop and, from a broader perspective, they are one step closer to the termination of the human species.

If this Genetic Use Restriction Technology (GURTs) were adopted everywhere on Earth, this would give transnational corporations like Monsanto absolute control over the world's food supply and thus indirectly over agricultural nations themselves. By the way, Monsanto's 2006 revenue was $7.6 billion. This corporate Goliath, which is the world leader in bioengeneering, produces among numerous other products the herbicide "Roundup," which some studies have associated with cancer. This controversial and litigious company also developed for the U.S. military the infamous chemical "Agent Orange," of which 12 million gallons were used during the Vietnam War and, according to Vietnam officials, killed or rendered seriously ill c. 500,000 people, not to mention the 500,000 children born with genetic defects; it also destroyed c. 14 percent of Vietnam's forest and 50 percent of mangrove areas.

More recently, Monsanto became embroiled in a lawsuit against the Canadian farmer Percy Schmeiser, who had his organic crop cross-contaminated from a neighboring field planted with Monsanto bioengineered crop. When Monsanto learned of this, the company aggressively sued Schmeiser for patent infringement. After several years of high-profile litigation, in

2004 the Supreme Court finally ruled 5:4 against Schmeiser.

Yet, the moral victory was Schmeiser's, because Monsanto was not awarded punitive damages and had to pay its own court fees. More importantly, organic farmers and agricultural organizations from around the world had rallied around him and contributed to his defense fund. In 1999, Schmeiser countersued Monsanto for $10 million for libel, trespass, and contamination of his fields with "Roundup Ready Canola." This case is scheduled to be heard in 2008.

Whatever the outcome of the countersuit will be, Schmeiser, who received the Mahatma Gandhi Award in 2000, has taught other farmers to stand up for themselves in the fight against bioengineered crops and pressures from agrichemical corporations like Monsanto.

Bioengineered crops are a potential threat to the food supply of future generations. It is always possible for genetically altered crops to succumb to pests or drought on a large scale, which would expose farmers and entire nations to a waiting period until the few corporations could provide them again with terminator seeds for next year's season. In the meantime, huge numbers of people would die of hunger, as would their livestock. Only a fiendish and greedy corporate mind could dream up such a potentially ruinous approach.

Understandably, local farmers in many parts of the world are outraged and consider this invasive technology as a form of "bio-piracy" and have been vigorously opposing it. As the Honduran peasant Rafael Alegría of Via Campesina, an organization representing over 10 million farmers from around the world, put it:

> Terminator [seed] is a direct assault on farmers and indigenous cultures and on food sovereignty. It threatens

the well-being of all rural people, primarily the very poorest."*

In 2000, the United Nations Convention on Biological Diversity recommended a moratorium on field testing and the commercial sale of so-called terminator seeds—a moratorium that, fortunately, was reaffirmed in 2006. Both India and Brazil with exception foresight have legislated to prohibit this kind of biotechnology. Other nations are poised to reconsider their present stance. Meanwhile, companies like Seed & Plant Sanctuary for Canada preserve organic "heirloom" seeds for future generations.

As an aside, we would like to make here some general comments about biotechnology, specifically cloning. In his book *Cloning the Buddha,* Richard Heinberg addressed the moral issues behind cloning, which had become a historical actuality in 1997 with the birth of the cloned sheep "Dolly." As he reviews the early development:

> The idea of cloning humans proved so disturbing to so many people that, by late 1997, President Clinton had convened an ethics panel to advise him on the subject; the FDA had ruled that any effort in that direction would require agency approval; and nineteen European nations had signed a treaty saying that cloning people violated human dignity and was misuse of science."**

Heinberg brings the full text of the "State of the World Forum Statement on Life and Evolution." The following excerpt is

---

*Quoted after the site "Ban Terminator" at www.banterminator.org

**Richard Heinberg, *Cloning the Buddha: The Moral Impact of Biotechnology* (Wheaton, Ill.: Quest Books, 1999), p. 10.

particularly relevant in the present context:

> Life must not be treated as a commodity that can be owned in whole or in part, by anyone, including those who wish to manipulate it in order to design new life forms for human convenience and profit. There should be no patents on organisms or their parts.*

Turning next to livestock "production," which globally claims c. 34 million square kilometers (c. 13 million square miles), we must note that the world consumes roughly 300 million tons of meat. That much meat eating also takes its toll on the environment in terms of soil degradation and water pollution, as well as the emission of greenhouse gases. To be more specific (citing 2002 statistics), the annual per-capita consumption of beef is 275 pounds in the United States and 238 pounds in Canada. The production of the former in carbon equivalents is 8,460 pounds; the latter, c. 7,380 pounds.

By comparison, China weighs in at roughly 115 pounds of meat per person per year, Nigeria at approximately 19 pounds, while the largely vegetarian nation of India still consumes 11.4 pounds, with meat eating becoming increasingly popular in the middle class.

In Chapter 2, we mentioned the detrimental effects of meat eating on water and the atmosphere. Every year, as noted previously, livestock is responsible for 18 percent of all greenhouse gas emissions, which total 24 billion tons of carbon dioxide equivalents. The digestive process in cattle produces some 100 million tons of methane annually, which is roughly twenty times more impactful than carbon dioxide when it comes to global warming. In the present context, however, we want to focus at

*Ibid., p. 233.

least briefly on the devastating impact of livestock on the land.

Livestock's main feed is corn, which explains why in the United States some 80 million acres are planted with this crop, yielding over 700 million tons annually.

In South America, 70 percent of deforested land is used for pasture land, while a large part of the remaining 30 percent is planted with feed crops. Here as elsewhere, livestock also is a significant threat to biodiversity. In other words, to raise and feed livestock, farmers destroy valuable wildlife habitat. This is part of what has been called livestock's "long shadow." So is overgrazing, which leads to soil compaction, erosion, and degradation.

To summarize, livestock production is catastrophic for the environment, because of the numbers involved. Again and again, in all these matters, we encounter the same basic problem, which is the size of the human population and its enormous demand on the Earth's resources.

Next we must turn to urbanization as another significant factor in the loss of arable land. According to the National Resource Inventory, in the decade between 1992 and 2002, the United States alone lost 8,900 square kilometers (2.2 million acres) of land to urban sprawl; much of it farm land.

Apart from decimating priceless arable land, urban sprawl is environmentally damaging also because it encourages an increase in automobile use due to commuting and transportation of goods and thus in greenhouse gas emissions. It also destroys wildlife habitat and displaces species.

In addition, because urban environments discourage walking and encourage overconsumption, they invite a higher incidence of obesity. Combined with air pollution, this makes for unhealthy living and costly healthcare.

Urban sprawl, moreover, involves extremely high costs in

maintaining roads and sewers, etc. As all municipalities know, this leaves little capital for creating and maintaining libraries, museums, parks, and so forth. It also generally undermines the spirit of community. Paved roads, parking lots, and shopping malls can hardly be said to be conducive to inner peace and harmony. Thus urban sprawl also clearly has an adverse effect on the mental state of city dwellers. It is scary to contemplate that humanity is becoming increasingly urbanized and that 80 percent of the population of some nations—like Canada—have become or will shortly become city dwellers.*

The intense shuttling of consumer goods from place to place is made possible by an elaborate grid of roads. In the United States alone, there are over 6 million kilometers of road; in China and Canada over 1.4 million kilometers each. According to a government statistic, in 2004 the United States had 243 million registered passenger vehicles, which is about 30 percent of all passenger vehicles in the world. The United States also has about 5 million large trucks. All this amounts to an enormous puff of greenhouse gases deposited in the atmosphere.

In 2002, the elaborate American road network had to cope with 11.7 billion tons of commercial freight. To this we must add freight by rail (1.9 billion tons), ships (1.6 billion tons), and airplanes (6 million tons), which exact their own price in terms of environmental pollution.**

This seems the right place to mention the 100-Mile Diet, which was initiated in 2005 by a Canadian couple, Alisa Smith and James MacKinnon of Vancouver. This started out as a simple personal experiment: Realizing that in North America our

---

*See, e.g., Douglas E. Morris, *It's a Sprawl World After All* (Gabriola Island, British Columbia: New Society Publishers, 2005).

**Statistics quoted after the U.S. Department of Transport's report "Freight in America: A New National Picture" (Washington, D.C., January 2006).

food travels an average of 1500 kilometers to get to us (and at a huge cost to the environment), Smith and MacKinnon committed for one year to only rely on food that was grown within a 100-mile radius. They managed to accomplish their goal at some personal sacrifice and time-consuming searches for local produce and products. Their blog at www.100milediet.org in which they reported their trials and tribulations achieved great popularity in a short span of time, and their book made the *San Francisco Chronicle* bestseller list.*

Automobile use for convenience and leisure is part of the modern definition of "personal freedom." Seldom do people think that this freedom could be illusory and actually constitute a form of thraldom to technology. In her book *Car Sick*, Lynn Sloman writes:

> In less than forty years, the car has become so intrinsic to the way we work, shop and spend our leisure time that it is almost inconceivable that we once managed without it. It is practically unimaginable that we might be able to use it less.**

Sloman, who quite reasonably favors a people- rather than car-centered society, argues that cars are not really the time savers we like to think but, in addition to polluting the environment, hog our time. This is so especially as long as we live in concrete jungles called cities, which favor automobiles instead of people.

Given the extent to which automobile use is gratuitous and

*See James MacKinnon and Alisa Smith, *The Hundred Mile Diet: A Year of Local Eating* (Toronto: Random House Canada, 2007).

**Lynn Sloman, *Car Sick: Solutions For Our Car-Addicted Culture* (White River Junction, Vermont: Chelsea Green, 2006), p. 11.

environmentally damaging, we ought to make it a moral consideration at this time of crisis. But as environmentalist and educator Michael Tobias observed, "[e]thics and alternatives are never popular."* The challenge, then, is simply to put responsibility before popularity and make the necessary lifestyle change.

We have talked about the misuse of land through agriculture, urban sprawl, and transport. A whole different form of land abuse are the countless garbage dumps, official and unofficial, that dot the landscape of industrial and industrializing nations.**

We know from megalopolises like New York what happens when urban garbage collectors go on strike, as has happened frequently over the years: Suddenly the streets are filled with literally tens of thousands of tons of garbage, which quickly become a significant health hazard.

Between 1948 and 2001, New York City generated enough garbage to create the environmental disaster known as Fresh Kills landfill on Staten Island, which is the largest human-made hill covering 2,200 acres and being taller than the Statue of Liberty. Every day, New Yorkers produce more than 12,000 tons of residential waste, which needs to be transported and for which land needs to be made available. As if to hide the evidence of its residents' profligate consumerism or unconscious behavior, New York has of late chosen to ship its garbage to other U.S. states and even to foreign nations.

Other developed nations follow a similar pattern. Landfills

---

*Michael Tobias, *After Eden: History, Ecology and Conscience* (San Diego, Calif.: Avant Books, 1985), p. 167.

**See, e.g., Elizabeth Royte, *Garbage Land: On the Secret Trail of Trash* (Boston, Mass.: Little, Brown & Co., 2005); Heather Rogers, *Gone Tomorrow: The Hidden Life of Garbage* (New York: New Press, 2005).

and waste transportation have become big business, which, moreover, often directly or indirectly seeks to undermine existing recycling programs or new recycling initiatives. The profitability of the waste management industry explains why, for instance, in the United States garbage is both exported and imported; curiously, in this odious trade, the 50 states have since 1997 ended up with a surplus of 48 million tons of garbage.

The waste disposal situation is similarly critical in other megacities. In 2006, when Athens's garbage collectors were ten days into a strike, more than 30,000 tons of rotting, malodorus garbage had piled up in the streets. According to the World Resources Institute, the United States generates c. 4.4 lbs. per person/per day and Canada 3.9 lbs. as opposed to Finland's 1.5 lbs. As could be expected, the two North American nations unimpressively lead the field. America's yearly 245 million tons of garbage is a telling symptom of the nation's penchant for overconsumption, instant gratification, and discardable lifestyle.

Turning to the waste itself, we know from a U.S. Environmental Protection Agency statistic that one third of all garbage is made up of paper, which is singularly reprehensible, given the preciousness of trees and the fact that paper can readily be recycled. Plastics account for nearly 12 percent of refuse.* This, too, is unforgivable and alarming, because plastic is a toxic substance that despoils land and water and takes on average thousands of years to disintegrate.

Annually, about 30 million tons of plastic resin are produced in the United States alone. This is the raw material for numberless plastic products—from construction materials to electronics, furniture, toys, agricultural products, medical gadgets, and so on.

---

*See Anthony L. Andrady, ed., *Plastics in the Environment* (Hoboken, N.J.: Wiley, 2003). See also Jeffrey L. Meikle, *American Plastic: A Cultural History* (Piscataway, N.J.: Routgers University Press, 1997)

The most trivial use of plastics is unquestionably in packaging (such as the infamous plastic shopping bag), which claims roughly one third of plastics usage. Over 80 percent of plastic items are discarded, even though many plastic articles could be reused or recycled. A significant portion of them, as noted earlier, ends up in the ocean where it wreaks major havoc.

We must highlight the fact that so-called "biodegradable" plastic could still consist of or include hydrocarbon-based toxic plastic. Now that the chemical industry has latched onto the identifiers "bio," "green," "eco," or "organic," consumers need to be especially careful and take the time to check out the actual ingredients in products carrying these labels. Of course, only because some consumer items are biodegradable or compostable does not mean that we now have a licence to continue our unsustainable lifestyle.

On the positive side, the U.S. national recycling rate of 32 percent for 2005 shows an improvement over previous years, and it prevented the release of c. 49 million tons of carbon dioxide into the atmosphere. Again, however, a great deal more could and should be done. Some cities and nations are better at recycling than others. Copenhagen, for instance, has been able to reduce the number of its landfills from 30 to 3 since 1991, recycling more than 50 percent of the city's commercial and industrial waste.

Ideally, in our view, all plastics production should be phased out quickly, though given the ubiquity of plastic items in our civilization, this is most unlikely to happen on a voluntary basis.

Wounding the Earth

Long before the invention of cities, humans dug into the ground to extract minerals—hematite for red pigment and flint for tools (both already in the Stone Age); stone for buildings; copper and quartz for tools and weapons; turquoise and lapis lazuli for jewelry, and so on.

Mining made the fateful event of the Industrial Revolution possible. Our technological civilization, in turn, has made mining into an extensive priority enterprise that continues to significantly alter the face of the Earth. The impact of mining on the environment is vast.* Even where environmental regulations are in place, mining still causes substantial damage in the form of soil erosion and the pollution of water and air.

Mining is also responsible for loss of biodiversity, which has become a grave issue around the world: As the diversity of life forms decreases, the survival of the remaining life forms is threatened. Great biodiversity is a product of hundreds of millions of years of evolution and a sign of health. Human civilization's modus operandi, which consists in the unrestrained exploitation of the natural environment, has had a devastating effect on biodiversity leading to the present soaring rate of species extinction.

Moreover, in all-too-many areas of the world, mining has directly contributed to human misery by displacing native peoples and ravaging sacred tribal land. To mention just one recent case: A corporation started to mine for gravel on land that the Mohawks had never ceded to Canada. Even as appointed Mohawk elders were negotiating with the government in Ottawa over land they already owned, the mining company was steadily

*See Fred G. Bell and Laurance J. Donnelly, *Mining and Its Impact on the Environment* (New York: Taylor & Francis, 2006).

removing an estimated 100,000 tons of gravel every year.

Wanting to put a stop to the mining while the predictably dragging negotiations were going on, a group of 150 protesting Mohawks took over the quarry. What they found to their utter dismay was large piles of junk created by mine workers over the years, which included toxic liquids and pieces of asphalt from a nearby highway. Whenever it rained, the liquids would run in a colorful stream downhill and pollute the groundwater.

Unless there is a local problem that affects us directly (such as toxicity or noise pollution), we seldom give another thought to the devastating impact of mining. Yet worldwide, mining is leaving hideous pockmarks on the environment and the human soul. Strip mining and mountaintop removal are particularly hideous and damaging to the environment.

The hunger for minerals (including metals) by industrialized nations is insatiable. According to a 1990 statistic, in that year the world mined 11 billion tons of stone, 9 billion tons of sand and gravel, 552 million tons of pig iron, 500 million tons of clays, 18.1 million tons of aluminum, and 8.92 million tons of copper, to mention just some of the most mined minerals.

In most instances, huge amounts of earth and rock have to be removed to obtain the ore. Today's brutal surface mining leaves behind a trail of devastation. Mountain tops, forests, agricultural fields are gobbled up by gigantic machines. Often all that is left behind is a veritable toxic wasteland.

Kennecott's Bingham Canyon copper mine is the largest human excavation in the world. Its open pit—a gaping wound that, as the company proudly announces, can be seen from the Space Shuttle—was created by removing well over 3 billion tons of material and spewing untold amounts of toxic gases into the atmosphere. Every day, another 450,000 tons of material are removed. The mine, which was started in 1906, has yielded about 17 million tons of copper. According to a report filed by Kennecott mistakenly in 1989, the mine had produced 130 million pounds of copper residue two years prior—a stupendous amount.

From our perspective, the most destructive branch of the mining industry is uranium mining.* The annual global uranium production is c. 40,000 tons, with Canada producing over 11,000 tons and Australia over 8,000 tons. Uranium is the heaviest of all naturally occurring elements, and its density suggests the enormous amounts of energy stored in this mineral. Uranium has 16 isotopes (variants) the most common of which are U-238 and the much rarer U-235. It is the latter isotope that is of immediate interest to the energy industry, because it packs more punch. By splitting the mineral's atoms (suitably converted into a gas) in a nuclear reactor through a controlled chain reaction, the mineral releases its hidden energy in the form of heat (translated into steam), which then drives huge electricity-producing turbines.

While natural uranium itself is only moderately radioactive (less than granite!), uranium ore is often found in conjunction with radium and bismuth, which are more strongly radioactive and present a potent hazard. The uranium mining tailings (debris) left behind, though considered only low-level radioactive, en mass represent a formidable and permanent threat. Just three uranium mines at the border between Germany and Czechoslovakia have produced radioactive tailings of more than 150 million tons. In other respects, uranium mines are environmentally just as destructive as any other mine.

Once uranium has been enriched in a reactor, it becomes extraordinarily potent as an energy source but also extremely lethal. Its industrial residue—nuclear waste—remains highly radio-

*See Broder G. Merkel and Andrea Hasche-Berger, eds., *Uranium in the Environment: Mining Impact and Consequences* (Berlin: Springer, 2005).

active for millennia. Plutonium 239, which is a radioisotope created in the nuclear process, remains lethal for 250,000 years. Thus far, no one knows how to dispose of nuclear waste *safely*.

In the past, several nations have been guilty of irresponsibly dumping nuclear waste in the ocean, and the damage done to the marine environment is inestimable. For lack of a better solution, some governments are burying their nuclear waste in deep shafts, while others are simply biding their time. In the meantime, in the United States alone over 50,000 tons of high-level radioactive waste are just sitting around in makeshift storage containers that will disintegrate long before their deadly contents.

The 440 or so nuclear reactors around the world have a combined output of more than 350,000 megawatts. To put this figure into perspective, New York must generate c. 36,000 megawatts to keep everything going but is under constant pressure to mobilize more energy for further growth.

It is easy to see how industrialized and industrializing nations, ever hungry for electricity, would find this technology highly attractive. Were it not for the radioactive tailings and the nuclear waste produced by reactors, it would indeed be a somewhat better alternative to coal- and gas-driven power stations. But, of course, we cannot remove radioactivity from the equation. Nor should we leave cleaning up radioactive waste, decommissioning nuclear reactors, and health care cost out of any cost effectiveness calculation: We would find that nuclear energy is unbelievably expensive.

We know from the disasters at Three Mile Island (1979) and Chernobyl (1986) that nuclear power involves an extreme risk. In 1985, the Nuclear Regulatory Commission testified before U.S. Congress that a severe nuclear accident has a 45-percent probability of happening in America within 20 years.

Such an accident is statistically overdue! The Chernobyl disaster released 50 million curies, which is fifty times more than the atom bombs dropped on Hiroshima and Nagasaki. Since radioactive dust particles were widely distributed by wind currents, some experts estimate that the Chernobyl accident might be responsible for up to 500,000 cancer deaths not only in the former Soviet Union but also elsewhere in the world.

Nor is the danger of this nuclear disaster over yet. Recently, scientists have concluded that the "Red Forest" near the exploded reactor and now consisting of dead pines represents a huge hazard. Desiccated trees can catch fire easily, and in this case the smoke released would again carry radioactive particles far and wide.

Despite the danger of inadequately stored nuclear waste and the great risk of theft by terrorists, who could easily construct a "dirty bomb," governments and of course profiteering corporations continue to promote nuclear power as *the* solution to humanity's energy problems and indirectly to global warming. Sadly, they have of late been joined by misguided spokespersons for the environmental movement.

As we see it, the use of nuclear power without a safe disposal method is humanity's greatest folly. In fact, we would argue that it amounts to insanity.

The Forest As Metaphor and Reality

The world is a forest. This may seem an odd statement considering our planet's forests are disappearing at lightning speed. What we mean to express by the forest metaphor is the Earth is brimming with life as only forests brim with life, at least until giant machines hack them down and with the subsequent action of wind and rain turn them into barren wasteland.

Each living organism is precious, each is playing its role in the total complexity of our biotic environment. Sadly, out of ignorance, indifference, and greed we are currently diminishing biodiversity at a rapid rate. As mentioned before, biologists are speaking of a Sixth Mass Extinction. **This is not just a neat label that you can file away and forget. It is a pressing *actuality,* as are pollution and global warming.**

A forest, to be sure, is not merely the sum total of trees in a given area. Even if we were able to see all the trees of a forest at once, we still would not see the forest. The reason is not that *forest* is simply an abstract concept, which it is as well. As a living reality, a forest is very much more than the trees that define its most notable feature. It is an incredibly complex living system. It is a living *whole.*

As such a forest is interconnected even with life forms that surround it. This is best illustrated by the remarkable discovery that there can be a reciprocal relationship between trees and salmon. While trees provide the necessary shelter and shade for salmon to spawn upstream, salmon in turn indirectly feed the forest with important nutrients. During their years of swimming in the deep waters of the Pacific Ocean, salmon store the heavy nitrogen N15 in their bodies. During the spawning season, many are caught by bears and dragged into the forest to be devoured. Their uneaten remains act as rich fertilizer.

In order to fully appreciate the following deliberations about forests and deforestation, we must don the philosopher's mantle for a moment. Facts and figures no doubt speak for themselves, but without the appropriate attitude they may not make their point. After all, we must expect for corporate bosses to know the same facts and figures, for they always manage to explain them away or talk around them when discussing the environment.

Especially when considering today's planetary crisis, we must not succumb to the convenient positivist habit of divorcing fact from value. Instead, we should permit the dreadful facts not only to inform our mind but impact our whole being. Only then will we find the wisdom and inner strength to respond appropriately.

From the perspective of global warming, forests are vitally important because they are first-rate "carbon sinks": Trees, soil, and peat absorb a considerable amount of carbon dioxide and the trees also give off oxygen. The entire complex process has a wonderful cooling effect, as anyone who has ever enjoyed walking in a forest knows. Forests are also rain producers.

In other words, we need our forests! While well-established forests tend to be carbon neutral, forests with lots of growing trees gobble up more carbon dioxide than they release at night. Old-growth forests are so important because they are a haven for an enormous variety of plants and animals. Unfortunately, it is the really large trees of ancient forests that the timber industry covets and mercilessly destroys.

According to The Global Forest Resources Assessment 2005, which studied 229 countries, the Earth's forests represent about 283 gigatons of carbon in biomass. That's 283 gigatons of carbon that did not end up in the atmosphere! Yet, because of deforestation, since 1990 this figure has been reduced alarmingly by 1.1 gigatons per year.

It is estimated that originally about half of the Earth's land surface was covered with forests. Today only one fifth of old-growth forest—some 15 million square kilometers—remains. Thirty-nine percent of this forested land is endangered by human activities (such as urban sprawl and especially logging). Since 1900 A.D., approximately 90 percent of West Africa's coastal rain forest has been destroyed; 65 percent of Latin America's; 93 percent of South Asia's, and an unbelievable 97 percent of Europe's forests. About 1700 A.D., North America was still covered with approximately 60 percent forest. By 1900, the forested area had shrunk to 45 percent. Today it stands at c. 28 percent. Logging is continuing unabated, and 900 million trees end up in American paper and pulp mills alone.

According to Greenpeace, planetwide industrial and illegal logging deprives the Earth of c. 23 million acres per year. Industrial logging yields an annual revenue of c. $150 billion. In addition, global industrial logging is supported annually by billions of dollars in government subsidies—an indicator of how badly governments want to keep their country's economy afloat.

It is virtually impossible to estimate the profits yielded from widespread illegal logging. The World Bank has estimated that it represents a loss of about $10–15 billion to the timber-producing countries. In Indonesia, an estimated 90 percent of logging is done illegally; in Brazil, it is between 60 and 80 percent. What makes this theft worse is the fact that the revenues from it have been shown to fund nefarious activities, such as civil wars and organized crime. All this legal and illegal profiting from forests is a minute fraction of what forests are "worth" in environmental terms if they were left standing instead of being ransacked.

Deforestation is a major factor in global warming. Amazingly enough, the Kyoto Agreement did not take deforestation into account in rating the participating nations' carbon emissions. The protection of forests should be made an absolute priority when considering global warming.

The devastation of deforestation can readily be seen in countries like Ethiopia, which half a century ago was still 98 percent forested but today is covered by only 14 percent. Similarly, China has lost most of its forests and only 18 or so percent of its land mass is still forested. According to Chinese government statistics, the nation has been planting roughly 35 billion trees since 1982. Able-bodied people are encouraged or even expected to plant several trees every year, which we think is an excellent policy worthy of emulation. And yet, every year 2,460 square kilometers of vegetated land succumb to desert, thus continuing the ecological downward slide. The nation's overly rapid development (ergo overlogging), mismanagement, and corruption all play a role in this fiasco.

Not too many years ago, I (Brenda) participated in a tree-planting program in Canada and over the years have planted thousands of trees. Tree planting is physically very demanding, because of the long, backbreaking hours involved in getting the seedlings into the ground. However, I found the experience emotionally extraordinarily rewarding. While I have not taken part in such a program recently, we both support Trees For the Future regularly and encourage others to do the same.

Unlike former U.S. vice president Al Gore, however, we do *not* rely on this and other similar programs for "offsetting" our carbon footprint. We know that we must also significantly reduce our carbon emissions by simplifying our lifestyle.

Our industrial civilization has an enormous appetite for trees, which are converted into thousands of consumer products, notably paper. The United States population makes up only 5 percent of the world's population; yet, according to the Natural Resources Defense Council, Americans use up 27 percent of all commercially harvested wood. This includes 80 percent of Canada's forest products, including lumber, newsprint, napkins, and toilet paper. Much of it comes from clearcutting virgin forest—a frivolous and inexcusable use of ancient trees.

Seventy percent of land animals live in forests, and as we deprive them of their natural home, we inexorably push them toward extinction. Moreover, some 60 million indigenous people depend on the forest for their subsistence, and another 1.6 billion for their livelihood. Deforestation, often in the form of ruthless clearcutting, is an unmitigated environmental and social disaster.

As we strip the Earth of forested land, we deprive ourselves of the very umbilical cord that helps us stay alive. Without forests, we are dead.

Of course, as with most other environmental issues, neither corporations nor conniving governments can solely be blamed for deforestation. Corporations log because there is a never-ending demand for wood and wood products. In the end, it is really the consumer who unthinkingly buys and also squanders forest products who is to a large extent responsible for deforestation. That's you and us.

To give just a few telling examples of overconsumption in this particular area: Paper consumption in the United States is phenomenal; annually and incredibly, Americans gobble up 87 million tons of paper and paper products. If every U.S. household were to replace a 500-sheet roll of toilet paper manufactured from wood cut from virgin forests with recycled paper, 297,000 trees could be saved, as well as 122 million gallons of water and 1.2 million cubic feet of landfill space. Also, in 1999, Americans used about 10 billion paper grocery bags, which translates into 15 million trees—a wasteful use of trees and an absurd number. "Paper or plastic?" The familiar question at the checkout stand in American supermarkets is altogether wrong. The answer should be "neither," and we all should bring our own shopping bag made from hemp or recycled material.

In March 2007, San Francisco passed a controversial bylaw prohibiting supermarkets and drugstores from using nonrecyclable and nonbiodegradable plastic bags made from fossil fuels. On April 2, 2007, the small town of Leaf Rapids in Manitoba, which was founded in 1970 as an environment-friendly town, went several steps farther. It became the first municipality in North America to ban single-use plastic shopping bags. The bylaw took effect immediately. All residents were given a cloth bag each, and retailers ignoring the new legislation can henceforth expect a $1000 fine per day. Bravo!

American businesses use a staggering 500 million wooden pallets, two-thirds of which are discarded on landfills without being recycled even once. The Chinese, again, produce some 45 billion disposable chopsticks every year, which consume about 25 million trees.

All these are most trivial applications of precious wood mostly from old-growth forests, which instead of being clearcut should be celebrated as a national and indeed international treasure. To stop this irresponsible, thoughtless decimation of the world's forests, we clearly need to transform the way we think about Earth's resources, followed by a radical lifestyle change.

Yoga practitioners, we feel, should have no difficulty extending their reverence for life to forests in particular. After all, many of the great masters who helped create the wonderful heritage of Yoga, were forest dwellers—sages like Vyasa, Vasishtha, Bharadvaja, Bhrigu, Markandeya, Uttanka, and so forth. Even the five Pandava princes, whose remarkable story is told in the *Mahabharata* epic (and in the *Bhagavad-Gita* contained in it), lived in concealment in the forest for twelve years.

The ancient sages saw the forest as an ally and took care not to inflict damage on it or harm its nonhuman inhabitants. The reverence shown to trees is captured in an ancient hymn of the *Rig-Veda* (10.146) addressed to the Spirit of the Forest (Aranyani), who is "rich in food [yet] uncultivated" and does no harm unless approached aggressively.

In the *Brihad-Aranyaka-Upanishad* (3.9.28.1ff.), which is widely held to be the earliest Sanskrit scriptures of its genre, we can find a highly instructive passage. It makes an analogy between the human being and a tree to illustrate the archaic teaching "As within, so without": Just as a tree has pulp, fibers, bark, leaves, sap, and seeds, a human being has a flesh, nerves, skin, hairs, blood, and semen. Unlike a tree, however, which does not continue to live and grow when cut down, a human being dies and then is reborn.

The gnostic teaching of rebirth, first distinctly enunciated in the early *Upanishads,* was in those days considered a great secret and hence had to be taught outside the village in the woods lest an unqualified person should overhear it.

This teaching forms an integral part of India's great religio-spiritual traditions of Hinduism, Buddhism, and Jainism and can also be found in many other traditions. According to a poll released by CNN in 1990, one third of Americans believe in reincarnation. A 1998 Gallup poll put the rate at one fourth, which is still very significant.

This is not the place to argue about the validity of this widely held belief. If true, however, then there is an interesting consequence: **We who are causing the present ecological crisis may well have to come back to harvest the karmic repercussions of our previous overconsumption, indifference, and lack of reverence for all life.**

5

HONORING THE BREATH OF LIFE

The One Breathing Breathlessly

Second to nitrogen, oxygen is the most common component of the Earth's atmosphere (nearly 30 percent). By mass, it makes up 86 percent of the world's ocean, and 49 percent of the Earth's crust. As a result of photosynthesis, phytoplankton produces roughly 70 percent of all oxygen on our planet, while vegetation, particularly trees, is responsible for the remaining 30 percent.

On average, a single mature tree can absorb c. 26 pounds of carbon dioxide (the bad stuff) in a year and in exchange generates approximately 260 pounds of oxygen, which is sufficient for a small child. According to NASA, at 1.85 pounds per day, an adult would need c. 675 pounds of oxygen per year and thus would require a little more than two mature trees or a rather large tree.

Unfortunately, as mentioned previously, we are mowing down trees as if they were blades of grass, and pollution is playing havoc with the tiny maritime organisms we call phytoplankton. To our knowledge, no one has yet precisely calculated what this devastation does to the production of oxygen on our planet. We think of air as another inexhaustible resource but evidently it is not. The air around us depends for its existence on chlorophyl-based organisms, which require a healthy environment for their continued survival.

This brings us to the Hatha-Yoga teaching that an adult individual takes 21,600 breaths every day, which corresponds fairly closely to scientific estimates. Assuming a yogically ideal life span of 84 years, this amounts to approximately 662 million breaths in one's lifetime. The Yoga masters apparently had no knowledge of the role of plant photosynthesis in the generation of oxygen or of the role of oxygen in human physiology. They did, however, recognize long ago the importance of breathing to physical health and, more importantly, to the mind-body connection.

Some five millennia ago, the archaic *Atharva-Veda* made the connection between life and breath, using the Sanskrit term *prana* for both. *Prana* means literally "breathing forth" (*pra* + *ana,* "to breathe"). On the one hand, it designates the air we breathe; on the other, it stands for the vital energy, the *élan vitale,* that is far more subtle than air and that is the primary life-giving and life-sustaining agent.

Prana is the omnipresent life force, which is found both within the body and all around it. Wherever life forms exist or are possible, *prana* is inevitably present. There is an essential interconnection between *prana* and the air we breathe. Good air, from a yogic point of view, is air rich in *prana;* bad air has a shortage of *prana.*

The pre-scientific notion of *prana* is rejected by mainstream science, though some intrepid researchers have adopted it to explain some of their findings, calling it "bioplasma" or "bioenergy."* This is not the place to go into details about this

*See, e.g., Hiroshi Motoyama, *Theories of the Chakras* (Wheaton, Ill.: Theosophical Publishing House, 1977); Hiroshi Motoyama with Rand Brown, (ctd.)

fascinating rediscovery of *prana*. Here we are primarily interested in Yoga's understanding and application of *prana* and what this means to practitioners today, who have to contend with air pollution.

Just as the water we drink and the food we eat can introduce toxic substances into the chemical laboratory of our body, so also does our breathing transport potentially polluted air into our lungs. Assuming that you are fortunate enough to be able to drink safe water and eat so-called "organic" food, you are still forced to breathe the same more or less polluted air as everyone else, which is a substantial health hazard.

If you live in or near a city, you can be sure that the air you breathe is contaminated by the exhaust fumes of thousands of automobiles and, perhaps, the smokestacks of factories or power plants. If you live in a rural area where crops are sprayed with pesticides, you are exposed to contaminated air as well. Even if you were to never leave your home, you would still have to deal with all the toxic outgasings of plastic materials, adhesives, and cleaning substances.

On average, North Americans spend up to 90 percent indoors. Apart from the lack of exercise this may indicate, indoor living also has a hidden factor that adversely affects many people's health. We are referring to the literally thousands of chemicals that can be found in products used in an ordinary household and, in particular, materials used in the construction of homes (composite wood products, glues, paints, synthetic materials like insulation and carpets, etc.).

This hidden factor has been identified as a significant health hazard only in recent decades. Largely out of ignorance,

---

Science and the Evolution of Consciousness: Chakras, Ki, and Psi (Brookline, Mass.: Autumn Press, 1978); see also Robert Becker and Gary Seldon, *The Body Electric* (New York: William Morrow, 1985).

our species has undermined even its most immediate environment—the houses we live in. As a result, we are suffering from debilitating headaches, nervous system diseases, and not least various types of cancer.*

Your safest bet would be to live outdoors under a tree in a protected grove, as many yogis of India have done for eons. We realize, this is neither universally possible nor practical. Alternatively, you can eliminate the worst air contaminants from your home and surround yourself with beautiful oxygen-producing plants or take daily walks in a treed garden or public park (providing it is big enough or not encircled by busy roads).

It is not difficult to see that polluted air not only affects the body adversely but also, via the nervous system and brain, can have a negative impact on the mind. To give a simple example: When we inhale air that contains a high percentage of carbon monoxide (caused by exhaust fumes, kerosene heaters, gas heaters, and coal or wood stoves, etc.), we quickly feel nauseous, dizzy, and disoriented, and may even faint. According to the World Health Organization, indoor air pollution claims the life of 1.6 million people every year. This is probably another gross underestimate.

Yogis have paid close attention to the link between breath and mind. When our mind is agitated, we tend to breathe unevenly and more quickly. When our mind is calm, our breathing tends to be harmonious and deep. Conversely, when our body is loaded with toxic chemicals, it strains to maintain its various functions by mobilizing the immune system. This physiological alert mode manifests in a more rapid heartbeat and an irregular and accelerated breathing rhythm. In turn, it throws the mind off

---

*For a closer look at the problems with new and old homes and possible solutions, see Athena Thompson, *Homes That Heal And Those That Don't* (Gabriola Island, B.C.: New Society Publishers, 2004).

balance. As within, so without.

It is no longer a question of whether or not the air we breathe is polluted but rather just how polluted it is. From NASA satellite photographs, we can tell the enormous extent of pollution in certain parts of the world, notably China, India, Europe, and North America. Our blue planet, which hangs like a rare jewel in the space of our solar system, is acquiring ever larger zones of ugly and noxious haze—a telltale sign of air contamination resulting from industry, automobiles, and livestock.

City dwellers are especially subject to the ill effects of polluted air. Anyone who has ever visited metropolises like New York, Los Angeles, Beijing, Mumbai (Bombay), Mexico City, Sao Paolo, or Tokyo will know what air pollution means: a brownish cloak hanging ominously over the city, watering eyes, dripping nose, sore throat, coughing, wheezing, headaches, and skin rashes.

Over the past decade, some of the big cities have made an effort to clean up the local air. They have succeeded only to a limited degree by tightening air pollution ordinances and also by planting tens of thousands of trees.

In 2007, New York's mayor Michael Bloomberg released the most detailed inventory yet on greenhouse gas emissions in U.S. history. Several months earlier, he announced a policy that would reduce emissions in New York by 30 percent by 2030. Considering the predicted repercussions of global warming, this does not seem much at all, but we have to remember that by then the city will have grown by another million residents to 9.2 million. Sometimes it feels to us as though the world were taking one step forward and two steps back.

According to Mayor Bloomberg's statistics, in 2005 the city of New York produced the carbon dioxide equivalent of 58 million metric tons, which amounts to about 1 percent of the

CO_2 equivalent generated by the United States as a whole. This corresponds to one third of the average per-capita amount. Probably the fact that New Yorkers are crammed into a relatively small space has something to do with this. Yet, according to the SustainLane U.S. City Rankings program, in terms of air quality New York comes in only at 42nd place among America's top fifty "green" cities.*

As part of this initiative, Mayor Bloomberg also announced that by 2017 he envisions to add 1 million trees to the 5.2 million trees already existing along city streets. Unquestionably, trees are life savers in such highly polluted environments. But residents must also participate by voluntarily reducing their CO_2 emissions through comparatively simple actions like car pooling, availing themselves of public transport whenever possible, purchasing and using fuel-efficient cars, limiting automobile use to essential purposes, and also buying products and produce that involve as little transportation as possible.

New York undertook its greenhouse gas emissions inventory as part of the Cities for Climate Protection campaign, which was launched by the International Council for Local Environmental Initiatives. ICLE, now called Local Governments for Sustainability, was founded in 1990 under the auspices of the United Nations Environment Programme.

No fewer than 750 cities are participating in this international campaign, including 240 U.S. cities. Among them is Los Angeles, whose air quality is the worst of the 50 ranked U.S. cities by SustainLane. Beijing, Mexico City, Mumbai, and Sao Paolo, like all too many other urban centers, have thus far failed to take incisive remedial action. With over 17 million residents,

---

*See Warren Karlenzig et al., *How Green Is Your City? The SustainLane US City Rankings* (Gabriola Island, British Columbia: New Society Publishers, 2007).

Sao Paolo is a veritable cesspit of pollution.

Several years ago, I (Georg) visited Brazil and remember developing a severe headache after spending just one day in that inferno of a metropolis. I was hosted by a friend, who kindly put me up in a room on the third floor of his building from where I could watch clouds of black smoke drifting by my window—the nasty-smelling exhaust fumes of a never-ending fleet of buses. By the end of my stay, which mercifully lasted only for one week, I was feeling thoroughly sick in body and at heart.

When still living in California in the late 1990s, I would periodically drive from Lake County through the Sacramento and San Joaquin Valleys to Los Angeles. I would then descend on the south flank of the Grapevine (officially the Castaic-Tejon Route) into the LA Valley where I would without exception face the unpleasant effects of heavily polluted city air on my respiratory system. My symptoms would promptly clear up once I had left the unhealthy air bubble of Los Angeles and found myself in Santa Barbara, all of 90 miles away.

Air pollution is particularly tragic in India, the birthplace of Yoga, which is busy trying to catch up with the rest of the industrialized world. Here is what journalist Edward Luce, who has spent a good many years in that country, says about India's air pollution:

> The quality of air and water in India is declining as rapidly as its economy is improving (without being factored in as a cost). It is estimated that one-eighth of India's premature deaths are caused by air pollution. . . . In the villages, people die young of respiratory diseases because they lack electricity or access to gas. The climate and the topography of the village make it impractical to burn cow dung, or wood, outdoors. So they light their fire

indoors , and gradually they are choked to death.*

Luce, who between 2001 and 2006 was stationed in New Delhi as a bureau chief for the *Financial Times,* failed to mention that New Delhi, India's capital, is considered by some the most polluted city in Asia. Beijing, with its over 11 million residents, cannot be far behind. By far the worst polluted city in the world, however, is Mexico City. Many of its 18 million residents are suffering not just from breathing difficulties but even permanent alterations of the cells of their nose and throat linings, foreshadowing possible cancer in later life.

In 2006, the World Health Organization (WHO) challenged the governments of the world to step up their endeavors to improve the air quality in cities. In its guidelines, formulated in consultation with 80 leading scientists, WHO specifically has singled out particulate matter (PM_{10}), sulfur dioxide, and nitrogen dioxide created by the burning of fossil fuels, as well as ozone.

Ozone is a protecting "sunscreen" in the upper stratosphere but is quite lethal when hovering close to the ground, as is the case in heavy traffic. In 1974, scientists confirmed that there was a huge and growing hole in the stratospheric ozone layer, which by 2000 had become a gulf of 28 million square kilometers (10.8 million square miles). Through this hole harsh ultraviolet rays have been pouring, threatening to alter and destroy the genetic materials of plants, animals, and humans. The phenomenal increase in the incidence of skin cancer is a direct consequence of the thinning ozone layer.

The ozone holes—one over the Arctic, the other over the Antarctic—are the surprise result of an escalating use of

*Edward Luce, *In Spite of the Gods: The Strange Rise of Modern India* (New York: Doubleday, 2007), p. 342.

fluorocarbons, as utilized in aerosol cans, refrigerators, and air conditioning units. With the phasing out of fluorocarbons in consumer products, some scientists are confident that the ozone holes will repair themselves; others are less optimistic.

In the meantime, ground-level ozone, which is generated by exhaust fumes interacting with other chemicals in the presence of sunlight, is continuing to compromise city-dweller's health. But since ozone, like other air pollutants, can drift for many miles, even those living in the suburbs or countryside can be exposed to its harmful effects.

If properly applied, the WHO guidelines would certainly help reduce deaths due to polluted air by 15 percent. According to Canadian government statistics, every year around 6,000 citizens die as a result of air pollution, and the figures for other countries are not dissimilar.

Apart from the ill effects of particulate matter, sulfur dioxide, nitrogen dioxide, and ozone, long-term exposure to air that has a high lead content from automobile exhaust fumes can also prove lethal. It is known to damage the liver, kidneys, and reproductive system, as well as impair brain functions—effects similar to those we mentioned in connection with water containing lead. Exposure to lead has also been linked with antisocial behavior, which has become a major problem in schools. According to the American Academy of Pediatrics, there is a distinct correlation between elevated levels of lead in the body and a lower level of intelligence.

It would appear that our cities breed not only unhealthy but also possibly mentally challenged citizens. But today, with the widespread use of pesticides, country air can be just as bad for your health. Living in Saskatchewan, which produces nearly half of Canada's wheat, we know that even wide open spaces are not necessarily healthy. They can be just as full of particulate matter as any large city.

Apart from burnings at waste disposal grounds, periodic burnings of crop, wood and coal fires for domestic use, the use of coal for generating electricity, this province also has its share of dust from wind-blown soil and road dust from numerous unpaved grid roads.

Besides, we are sorry to say, Saskatchewan is the highest CO_2 producer in Canada because of its fossil fuel production, reliance on coal for electricity generation, and intensive transportation (most of its natural resources and consumer products are exported to the United States or still farther afield).

In Chapter 3, we already mentioned air pollution due to intensive and extensive hog farming. Other livestock exacts its own, comparable price in terms of air contamination. Recognizing this problem, the Sierra Club has recommended a ban on permitting new open-air lagoons (for liquid manure) and a program of phasing out existing installations. It also has argued that intensive factory farms should be treated as all other industrial polluters and be subjected to far more stringent standards in order to help drastically reduce the generation of greenhouse gases.

Locally, perhaps the most lethal part of air pollution in Saskatchewan and other grain-producing regions is the widespread use of pesticides—some 8 million tons every year!* Even though this is only 50 percent of the rate of pesticide use in the United States and 20 percent of that of France, it still amounts to a weighty 8 tons per Saskatchewanian. From our

*We must add the following qualifier here: Since pollution recognizes no national or ethnic boundaries, it spreads everywhere, and so pesticides used in one area can also be found in other areas, even where they have never been used. This is the case with the nerve toxin Malathion found in the breastmilk of Eskimo women.

perspective, this is 8 tons too many.

Why, then, do we choose to live here? Well, there are few places in Canada and the rest of the world that do *not* have some form of air pollution or where other conditions (such as the sociopolitical landscape and quality of living) are equal. Fortunately, Saskatchewan comprises just over 650,000 square kilometers and is sparsely populated. Almost half the land is (still) forested, with 60,000 square kilometers taken up by waterways. For most parts of the year and in many areas even the wide-open Prairies are still a welcome refuge from degraded air. Besides, we do not have to put up with block after block of concrete buildings, dodge rush hour traffic, or purchase our food from supermarkets whose goods have a high environmental price tag.

The Brewing Storm

Carbon dioxide, roughly 80 percent of which is generated by cities, has been identified as the main cause of global warming, and global warming is finally acknowledged as a reality—even by the anti-environmental Bush Administration.* What does this mean in concrete terms?

The Sun constantly bombards our planet with all sorts of radiation—from invisible infrared light to visible light to invisible but lethal gamma rays. A significant portion of this spectrum of rays, as we all know, is made up of infrared light, which we experience as heat. Our solar system's star bestows life in the

*George W. Bush, Jr.'s environmental policies patently favor industry, and many environmentalists would typify them as singularly hostile to Nature and those championing Nature's cause.

form of warmth and light. But for life to be possible, there also needs to be an atmosphere.

For millions of years, the Earth's biosphere has been made possible by a finely balanced and multilayered mantle of air extending around the entire globe. As far as scientists can tell, certainly for the past 10,000 years our planet has—until recently—enjoyed a basic thermostat setting of 14° Celsius (or 57.2° Fahrenheit) at the surface.

As the atmosphere fills up with a surfeit of carbon dioxide and other pollutants that are entirely due to human activity, the atmospheric mantle is getting thicker. Greenhouse gases contribute to capturing heat in the lower regions of our atmospheric envelope. This leads to an expansion of the air, which then heats up the ground and the ocean, which, in turn, contributes to further warming of the air. And so on.

Scientific studies have conclusively shown that our ailing planet's basal temperature has risen 0.4° Celsius in just a single century. This may seem a laughably minute amount, but it is not, as we can tell from the signs of global warming, which are all around us. What is more, now that an imbalance has been triggered, this so-called Greenhouse Effect can be expected to continue and even get worse for several centuries until a basic balance is restored. Even if all pollution were to miraculously stop today, the Earth would continue to heat up.

Most recently, scientists have discovered an added complication to global warming. This is the paradoxical effect of what they call "global dimming." This term refers to the fact that there has been a gradual reduction of sunlight due to particulate matter in the stratosphere. This, in fact, acts as a damper on global warming. As we clean up the air, more sunlight hits the surface of the Earth, the air trapped between the ground and the CO_2 layer in the stratosphere heats up even further.

Global dimming was discovered in the wake of the September 11, 2001, terrorist attack on the twin towers in New York City. The U.S. government banned all commercial and private air travel over the country for three days. As a result, the skies cleared up and the sun shone brightly. Then David Travis, a scientist working for the University of Wisconsin, discovered that in that short span of time the change in temperature range was 1° C! He realized that the crisscrossing contrails produced by jet planes that normally fill the sky had acted as a mirror allowing less sunlight to reach the ground. This important discovery is surely the only good thing to come out of 9/11!

This seems the place to interpolate a brief consideration about jet planes and air travel. According to the Office of the Assistant Secretary for Aviation and International Affairs, in the 12 months ending September 2006, 152.9 million passengers were transported on U.S. and foreign air carriers between America and the rest of the world. This represents 54.9 percent of worldwide passenger transportation by air.

Despite the fact that the news about global warming has been widely disseminated, U.S. air travel is up by 2.6 percent from the preceding year, and this upward trend is expected to continue. A news release issued by OAG (Official Airline Guide) on May 8, 2007, mentions that there are 114,000 more flights and 17.7 million more seats in 2007 than in 2006. The main reason for this increase is 12 million more low-cost seats offered by various airlines.

In fact, air transportation is the fastest growing source of carbon dioxide pollution! The exhaust (contrail) left behind by high-flying jet planes consists not only of water vapor but also unburned hydrocarbons, sulfate, nitrogen oxide, carbon dioxide, and not least particulate matter responsible for global dimming. A flight from New York to Denver produces 840 to

1,660 pounds of CO_2 per passenger! This is roughly the carbon dioxide emission of an SUV over an entire month. In other words, jet travel is contributing to the Greenhouse Effect. In an article published in *USA Today,* Gary Stoller states dryly, "Aviation and the environment are on a collision course."*

Global warming and global dimming come as a package deal and cannot be addressed separately. In down-to-earth practical terms, the only way to avoid total disaster is to not merely reduce emissions that increase particulate matter in the air but also to drastically reduce carbon dioxide emissions. This is why Al Gore's simplistic "solution" to global warming will not work and will indeed have the opposite effect of what is intended.

The latest climate change models predict a global temperature increase of up to 5.8° Celsius leading to a basal temperature of 19.8° Celsius (or 67.6° Fahrenheit) by 2100.** A significant rise in temperature would have catastrophic consequences. Some estimates consider a 9° Celsius increase possible by the end of this century, which would also cause the ocean to warm to a point that could be fatal not only for marine life but all life on Earth.

The British climatologist Peter Cox, who thinks the standard climate models are wrong, calculated that a warming by 10° Celsius could release the 10,000 billion tons of methane gas that are currently resting at the bottom of the ocean in frozen form. Now, methane is eight times stronger than carbon dioxide. As a consequence, it would return the Earth to the kind of boiling

---

*Gary Stoller, "Concern grows over pollution from jets," *USA Today,* December 19, 2006.

**See the scientific monograph by Mohan Munasinghe and Rob Swart, *Primer On Climate Change And Sustainable Development: Facts, Policy Analysis and Applications* (Cambridge, England: Cambridge University Press, 2005), p. 27.

cauldron it was 4 billion years ago!

In any case, however moderate global warming might turn out to be, it still will have far-reaching and highly undesirable worldwide repercussions. These include rapid melting of the ice caps, polar caps, and glaciers, a rise in sea levels, more frequent and intense heat waves and droughts, devastating mega-hurricanes, and the displacement or extinction of many more species than is the case already.

Here are some examples to render these abstract concepts more concrete: In 1995, a 2,000-square-kilometer chunk of ice broke off the Larsen Ice Shelf in Antarctica, "one of the most spectacular and nightmarish manifestations yet of the ominous changes occurring on the planet."* Seven years later, an even bigger slab disintegrated, releasing 720 billion tons of ice into the Weddell Sea. In 2003, this once enormous shelf of 11,512 square kilometers was measuring a mere 2,667 square kilometers. Many of the other 14 large ice shelves of the Antarctic have been receding at an alarming rate.

The Arctic ice is also shrinking dramatically and at the present rate could be ice free by 2100. While the temperature in the Arctic is rising twice as fast as elsewhere on the planet, in the Antarctic the rate is five times the global average. Climatologists are watching Greenland with particular worry, because, Greenland's coastal ice is melting at the rate of 11 cubic miles every year. At the time of writing, the most recent NASA report (dated May 29, 2007), based on satellite sensor readings, revealed that the last snow melt occurred ten days longer than the average since 1988.

This melt-down has already contributed significantly to rising sea levels. If all of Greenland's ice sheet were to thaw, the

*Ross Gelbspan, *The Heat Is On: The High Stakes Battle Over Earth's Threatened Climate* (Reading, Mass.: Addison-Wesley, 1997), p. 1.

level of the ocean would increase by about 20 feet.

Global warming is perhaps most dramatically demonstrated in the rapid melting of glaciers over the past century. In 1991, the news media featured the discovery of a 5,300-year-old "ice man" dubbed Ötzi found in a glacier of the Austrian Alps. Fascinating for archaeology buffs, this discovery also signalled that this particular glacier had retreated to its position of 5,300 years ago.

As studies have shown, the world's glaciers are shriveling up fast. For instance, the Himalayan glaciers, which are the largest body of ice after the polar caps, cover a vast area of 34,000 square kilometers. In the Nepalese part of this sweeping mountain range alone there are c. 3,300 glaciers, with 2,300 having glacial lakes. Annually, the Himalayas supply 8.6 million cubic kilometers of water. What would happen in the valleys if the glacial lakes were to burst their banks or if the glaciers were to vanish? This is not just a hypothetical question, because flash flooding has already occurred repeatedly during the past seven decades, claiming human and nonhuman lives and causing tremendous damage to property and the environment.

More importantly, glaciologists and hydrologists have expressed their concern that many of the glacier-fed big and small rivers of Northern India might run dry as early as 2035. This would leave several hundred million people settled in Northern India without water. Few of them are aware of the problem, partly because of lack of education and partly because the Himalayan glaciers are not adequately monitored and few studies are published.

Another telltale sign of global warming are rising sea levels. During the last 100 years, the ocean has risen by about 9 inches. Again, this does not seem much. But island nations like Tuvalu in the South Pacific know better. The island of Tuvalu is a 10-foot-high atoll island, which is now experiencing regular flooding, causing food crops to be destroyed and land to be eroded. Sometimes children are forced to swim to school, because the road is flooded.

In 2004, in recognition of the threat to their continued existence, forty-three island countries from around the globe formed the Alliance of Small Island States to ponder their fate and find solutions for rising ocean levels and increasingly injurious weather patterns. More than two thirds of the world's settlements harboring well over 600 million people are along coastal stretches. Four decades ago, about 7 million people were subject to flooding annually; now this is the disruptive experience of 150 million people.

Half of the surface area of The Netherlands is only c. 3 feet above sea level, and 27 percent of it is below sea level. The latter part is inhabited by 9.4 million people, or 60 percent of The Netherland's population. Over 2,000 years ago, the Friesians constructed the first dams to brace against the North Sea. In 1287, the dykes collapsed and the country was flooded. Between 1927 and 1932, the 30.5-kilometer-long Barrier Dyke was built, but it is not expected to stem the rising tides.

In preparation of the predicted further climb in sea level, The Netherlands have been experimenting with a variety of technological solutions to the problem of inundation, such as amphibious homes and buoyant roads. Few, if any, of the island nations could afford to go this high-tech route.

Under the impact of rising temperatures of air and land, as well as the ocean, the world's climate has become destabilized. Droughts occur more frequently in already dry areas (such as Australia, which is fast running out of water); places that have never had snow fall in historical memory or only moderate snowfall are suddenly covered with snow (in 2006, several

Japanese towns had to be dug out from under a thick blanket of snow; torrential rains surprise people in locations that have had an average amount of precipitation (such as New Zealand in 2007).

The most devastating harbingers of global warming are mega-hurricanes like Mitch (1998), which killed 10,000 people and robbed 3 million of their homes in the Caribbean, and Katrina (2005), which devastated New Orleans and has been called the "deadliest and costliest" hurricane ever.

Long-term forecasts warn of more and even bigger hurricanes, which will take their toll on agricultural land, cities, economy, and not least human lives. The tropics are particularly affected by this. On April 6, the United Nations released a report compiled by a panel of experts and intended for the world's policy makers. With a degree of "very high confidence," it warned that even a temperature rise of 1° Celsius would imperil 30 percent of species. It also would lead to further water shortages, droughts, and flooding.

Global warming, to be sure, will play a mounting and increasingly interfering role in modern civilization's life. Some observers are arguing that it may yet spell the end of human civilization as we know it. "A recent peer-reviewed study by 11 prominent researchers," states Ross Gelbspan, "found that unless the world gets half of its energy from non-carbon sources by the year 2018, the planet will see a quadrupling of atmospheric carbon levels—which would clearly be catastrophic."*

Remarkably and sadly, the United States, which is the largest consumer country in the world and responsible for 25 percent of global greenhouse gas emissions, has consistently pursued

---

*Ross Gelbspan, Foreword to Guy Dauncey with Patrick Mazza, *Stormy Weather: 101 Solutions to Gobal Climate Change* (Gabriola Island, British Columbia: New Society Publishers, 2001), p. xi.

environmental policies that almost ensure that the worst-case scenarios worked out for global warming will come true. Here are our principal reasons for making this claim.

In 1994, the nations of the world came together to consider "climate change" as a consequence of the Greenhouse Effect and agreed to reduce CO_2 emissions to the 1990 baseline level in an effort to slow down global warming. At that time, not every participant nation concurred that there was sufficient scientific evidence for global warming.

The United States, the biggest consumer country on Earth, has been among those nations that are still dragging their heels. Millions of people had hoped that the U.S.A, as the world's leading nation from a political and economic perspective, would play an exemplary role at the 1997 Kyoto meeting. To this day, America has not ratified the Kyoto Protocol, along with Australia, Monaco, and Liechtenstein. The juggernauts China and India are large-scale environmental disasters in the making.

In June of 2007, the Group of Eight (G8) met for what was supposed to be an important summit on climate change. The Group of Eight is made up of the eight wealthiest countries on Earth—Canada, France, Germany, Great Britain, Italy, Japan, Russia, and the United States. Germany's chancellor Angela Merkel, who presided over the meeting, had expressed her pessimism about the summit's success largely because of the U.S. government's recalcitrant stance. She was proven right. The G8 nations failed to achieve any firm commitment to reducing greenhouse gas emissions rapidly and significantly.

This is also the basic problem with former American vice-president Al Gore's message about global warming, as epitomized in his Life Earth Concert, which was directed at the younger generation. Gore's recommendations fail to take into account the phenomenon of global dimming. Nor has Gore thus far squarely faced the fact that the present rate of consumption in the United States and other developed nations simply cannot be "offset." Presumably, any suggestion that we must reduce our level of consumption would be too unpopular a message and also get him into immediate conflict with the corporate world.

When confronted, Gore shrugged off his own extravagant lifestyle with the remark that everyone's ecological footprint is different but that he did his best to offset his footprint with so-called carbon credits and by participating in CO_2-reducing programs, such as tree planting.

We readily concede that a former U.S. vice-president's duties may involve extraordinary social obligations and functions, but this does not exonerate him from the now obvious and primary obligation of simplifying his or her lifestyle. **Offsetting a rather large ecological footprint is certainly commendable, but it is not the same as living responsibly and reducing one's footprint to a sensible size.** We all need to take a hard look at our own footprint and identify those areas in which we can make big and small improvements.

Gore's orientation, we feel, might prompt others to likewise continue with their overconsumption on the grounds that they, too, can somehow offset it. This appears to be the current mood among his Hollywood supporters. Such an attitude is in our view irresponsible and damaging. You cannot merrily drive around in an SUV and clock thousands of miles in air travel *and* be ecologically responsible, whatever else you might do for the environment. This kind of wrong thinking is our only reason for highlighting Gore's approach here.

In the midst of the public euphoria over Gore's environmental campaign, we were glad to discover a few discerning voices, notably the Transnational Institute. In February 2007, the Institute released a report authored by Kevin Smith entitled "The

Carbon Neutral Myth: Offset Indulgences For Your Climate Sins." The report is highly critical of any approach that seeks to deal with global warming solely through "carbon offsetting." In Smith's words:

> The sale of offset indulgences is a dead-end detour off the path of action required in the face of climate change. There is an urgent need to return to political organising for a wider, societal transition to a low carbon economy, while simultaneously taking direct responsibility for educing our personal emissions. . . It is hoped that the rising awareness of the shortcomings of offset credits will contribute to a reformation of the climate change debate.*

Of course, switching to green energy is a wonderful thing. Also, planting trees around the world is a good way to reduce CO_2 pollution, and we vigorously participate in such a program ourselves.** But planting even millions of trees will never cancel out the pollution damage wreaked by the current lifestyle to which the citizens of developed nations have become accustomed.

Another disquieting factor about Gore's global warming campaign is that in his recommendations for cutting greenhouse gas emissions, he seem to totally ignore the considerable carbon dioxide and methane production by the livestock industry and the baneful environmental consequences of meat consumption

---

*Cited from the PDF version of Kevin Smith's report published by the Transnational Institute, p. 7. The version is available online at www.carbontrade-watch.org/pubs/carbon_neutral_myth.pdf.

**The program we currently support is Trees for the Future, located online at www.treesftf.org.

in general.

While we do not wish to attribute to Al Gore any sinister motivations, we would nevertheless argue that his overall position is misleading and possibly overly political. It also still reflects America's basically self-indulgent cavalier attitude toward the environmental crisis. **There is no quick fix for global warming other than to learn to tread lightly upon the Earth. And we must learn this lesson very, very quickly. The future of humanity and of most, if not all, nonhuman life forms depends on it.**

The interconnectedness of life is nowhere more apparent than in the air we breathe. Spencer R. Weart, the director of the Center for the History of Physics, which is part of the American Institute of Physics, has come up with an interesting calculation.* According to him, it takes just one month for the air (CO_2) we exhale in any given moment to be completely dispersed over the globe. In his well-argued book *Earth in Balance,* Al Gore put it graphically (and correctly):

> In every breath we take, we bathe our lungs in a homogeneous sample of that same air—many trillions of molecules of it—with at least a few in each breath that were also breathed by Buddha at some point during his life, and a like number that were breathed by Jesus, Moses, Mohammed—as well as Hitler, Stalin, and Genghis Khan.*

Given the importance of the breath in Yoga practice, this fact should make us particularly thoughtful. We cannot avoid the moral issues arising from our fundamental interconnectedness

---

*Al Gore, *Earth In the Balance: Ecology and the Human Spirit* (New York: Houghton Mifflin, 1992), p. 84.p. 84

with our physical environment and all life. What we do or don't do is relevant not only for us personally but can have significant repercussions for our biosphere as a whole.

Smoking tobacco is not only an unpleasant experience for nearby nonsmokers, as we quickly discover when we abandon the habit of smoking, it is also a health hazard for ourselves and others around us. Driving an automobile for no *really* good reason or jet-setting across the world is an even worse habit, which adversely affects us and many others. Allowing trees, which extract carbon dioxide from the air, to be felled by the millions is equally reprehensible from a moral perspective. So is our mostly wasteful use of nonrecycled paper or our tacit support of book and magazine publishers that out of lethargy or greed refuse to switch over to environment-friendly paper and methods of printing. And so on.

Air pollution is the principal cause of global warming, and global warming has been identified as the greatest threat to humanity's survival. This should be enough of an incentive to change our ways *radically*.

As Yoga practitioners, we must of course also bear in mind all those countless fellow beings who are the victims of human-caused (anthropogenic) environmental devastation. By the billions, they are being forced out of their habitats and condemned to death. We cannot allow ourselves the luxury of thinking that those individual beings—fish, insects, birds, land animals—are unimportant. They *all* have a right to enjoy life.

The loss of any species as a result of human ignorance, indifference, and avarice is unacceptable. We should weep for all those species that we have already driven into oblivion or that are teetering on the brink of extinction. Then we should wipe our tears and seriously commit to restoring our biosphere.

Let us be clear: With every species becoming extinct,

humanity moves inexorably closer to its own grave. After us, there will be no one to weep over our demise.

In February 2007, a panel of scientists representing 113 nations, issued a rather bleak report on climate change. It left no doubt about the fact that humanity is the cause of the worldwide climatic shifts that are occurring and threatening the survival of our species and possibly all living being on our planet.

Realizing the severity of the threat of global warming, some governments have begun to initiate corrective programs, but without the United States, Canada, China, and India coming aboard wholeheartedly, this rescue mission is unlikely to succeed in any significant way. Nevertheless, we all must do what we can to stave off escalating disaster.

Pollution is not merely an objective fact that we can quote and turn into statistical probabilities. Rather, it is a value-laden fact. Like overconsumption, which is one of the causes of environmental degradation, the pollution of air, water, and soil has profound moral implications (and karmic consequences) for every human alive.

Each of us is both a cause and a victim of pollution. We cannot hope to rectify our status as victims of pollution if we fail to change our lifestyle, if we fail to transform ourselves.

As Clive Doucet, poet and member of the Ottawa City Council, put it well:

> Will the planet give us enough time to get it right? I don't know, but what I have learned as a city councilor is clear and unequivocal. There is nothing does doesn't connect. We are connected from the division of the first cell that ultimately creates a human being through to our shared history of cities and human civilization, to

our mutual dependency on the planet's biosphere. To succeed today, we must remember how essential it is to make those connections work; and conversely we must remember what happens when we fail to connect the dots.*

As we argued earlier, environmental pollution has its correlate in our mind. As within, so without. We must transmute our thinking (our values, beliefs, and attitudes), so that we can properly appraise the damage we have done to the Earth, our only habitat, and also see very clearly what we must now do to correct what amounts to a traumatic situation.

Already back in 1992, some 1,700 leading scientists from around the world issued a memorandum warning governments and the public that "[h]uman beings and the natural world are on a collision course" and calling for fundamental changes. "No more than one or a few decades remain before the chance to avert the threats we now confront will be lost and the prospects for humanity immeasurably diminished. . . . A new ethic is required—a new attitude towards discharging our responsible for caring for ourselves and for the earth."**

Of all the groups of people we can think of, none is more qualified and also more obligated to step forward to be counted than the group of spiritually committed individuals. This includes those who are engaging Yoga as a spiritual discipline—the sincere practitioners of Hindu, Buddhist, and Jaina Yoga. Even though they are a small minority within the large contemporary

---

*Clive Doucet, *Urban Meltdown: Cities, Climate Change and Politics As Usual* (Gabriola Island, B.C.: New Society Publishers, 2007), p. 212.

**Cited on www.ucsusa.org/ucs/about/1992-world-scientists-warning-to-humanity.html.

Yoga movement, they still number in the tens of thousands.

But, naturally, we are hoping that those who have adopted Yoga primarily for health and fitness reasons will likewise realize the utmost necessity of switching to a green lifestyle. It is our contention that a green lifestyle is implicit in traditional Yoga. On our website (www.traditionalyogastudies.com), we feature the motto "The color of Yoga is green." This is meant to underscore the fact that if we follow the traditional teachings of Yoga, we are inevitably led to an environment-friendly lifestyle. Conversely, a green lifestyle meets the spiritual heritage of Yoga at least halfway.

To end with a quote from Sri Aurobindo, who, more than any other modern adept of Yoga, favored an integral orientation:

> The affirmation of a divine life upon earth and an immortal sense in mortal existence can have no base unless we recognise not only eternal Spirit as the inhabitant of this bodily mansion, the wearer of this mutable robe, but accept Matter of which it is made, as a fit and noble material out of which He weaves constantly His garbs, builds recurrently the unending series of His mansions.*

In plain, nontheological language: the world is not merely a meaningless and random conglomeration of atoms but a highly organized dynamic web of energies that allows consciousness to express itself creatively. Yoga is about discovering the true nature of consciousness (or mind) and of matter. Thus, Yoga is intrinsically reverential toward the mind, the body, and the world at large.

---

*Sri Aurobindo, *The Life Divine* (Pondicherry: Sri Aurobindo Ashram, 10th ed. 1977), vol. 1, p. 6.

6

GREEN YOGA ACTIVISM: WHAT YOU CAN DO

We have found that most people lack adequate information about the state of the world in which we live. Throughout this book, we have given you numerous statistics to help you better understand the problems we are facing today, but no book on the environment would be complete without a list of practical solutions that will help you initiate positive change. We certainly have created the huge mess we are in; but we also have the ability to change things around. We understand that global transformation is neither easy nor instant. *But* in regard to the environment, personal change can be relatively easy and can happen immediately.

Are we challenging you? You bet we are! We are hoping that you will take responsibility for your own actions and start doing as much as you possibly can today, *right now*. Nothing would thrill us more than to hear that everyone who read this book decided to further educate himself or herself and then become a spiritually minded environmental activist, or Green Yoga activist.

We make the optimistic assumption that everyone who reads this book is either already environmentally active or is ready to begin. To make things more convenient for you, we have listed a selection of websites, documentaries, online documents, and books and also have furnished you with an initial list of practical things you can do. In compiling this list, we have taken into account that finances may limit you from taking bigger steps and therefore have made sure to also include various practical ideas that cost absolutely nothing but are still very effective.

The Internet is full of resources (including organizations) that will assist you in finding yet more ways to actively protect and restore our planet. We suggest that you make a point of checking out the links page on websites, because they are bound to direct you to other helpful sites. We, furthermore, recommend that you create or participate in a local action group, which will help each member stay focused on the challenge of making his or her whole lifestyle green. Our free e-newsletter *Green Yoga Initiative* provides lots of practical help to Green Yoga activists. Moreover, we are putting together a *Green Yoga Workbook* for those who find it useful to have step-by-step guidance in personal change and global transformation.

Practical things that you can do to make a significant difference

- **Simplify your life as much possible.** Become aware of how your lifestyle impacts the environment. Make a list of what kind of "stuff" you use daily, weekly and occasionally as well as how you live your life in general. This practice may be mind-boggling at first but it will definitely help you identify all those areas in which positive changes can be made.
- **Think twice before you purchase.** Do you really need it? Can you find it used? If it is broken, can it be repaired?
- **Reduce, reuse, and recycle!** Try to reduce waste by choosing

reusable and recyclable products. Remember that a product may say it is recyclable but you may not be able to recycle where you live. Learn about what your community recycles. The United States generates approximately 230 million tons of "trash" annually of which 70 percent is recyclable or reusable materials BUT only about one quarter of it is actually reused or recycled.

- **Try to find products with as little packaging as possible**. A perfect example is a box of tea that has tea leaves in a bag that is wrapped in paper or plastic that is again boxed and then wrapped in plastic. Is this overpackaging or what?
- **If you are looking to replace a major appliance, choose an energy efficient one with the Energy Star label on it.** A new Energy Star-approved refrigerator, for example, uses up to 40% less energy than models made prior to 1993.
- **Purchase only 100% recycled and chlorine-free paper products or try an alternative paper like hemp or kenaf.** This includes printing and writing paper, paper towels, toilet paper, and tissue paper. According to the National Resources Defense Council, if every household in the United States were to replace just one roll of virgin fiber paper towels (70 sheets) with 100% recycled ones, we could save 544,000 trees.
- **Take a shower instead of a bath and challenge yourself to make it the shortest shower possible.** Water for bathing and showering accounts for two-thirds of all water-heating costs. Besides, most areas nowadays are suffering from water shortage. This is one of those ideas that does not cost you penny, but could save you dollars.
- **Install low-flow shower heads and low-flow faucet aerators.** Low-flow showerheads use about 2 gallons of water per minute compared to conventional showerheads that use

4-7 gallons per minute. This low-cost idea has significant savings for water, energy, and your pocketbook.

- **Switch your washing machine to cold water wash only**. If all U.S. washers were switched from hot water to cold water, it could mean a saving of about 30 million tons of CO_2 per year.
- **Use an old-fashioned clothesline or clothes-drying rack instead of your dryer whenever possible**. Make use of the gift of free energy from the wind and sun.
- **Turn down your water heater to 120° F or 49° C.** Most water heaters are set to have water scalding hot, for which there is no need.
- **Repair any leaks in your faucets and water pipes**. A typical steadily running leak can waste over 40 gallons or 151 liters of water a day!
- **Turn your thermostat down at least 2° down in the winter and 2° up in the summer**. You could save up to 2000 pounds of carbon dioxide a year just with this simple procedure.
- **Turn off every appliance that you are not using and unplug it at night or when you don't expect to use it for a longer stretch of time**. A computer, for instance, uses about 15 watts of power when plugged in but is not in use, and even a microwave oven uses 8 watts of power when not in use! Simply get into the habit of unplugging electric equipment when you are finished with it.
- **Purchase locally grown and produced organic food whenever possible or better yet, if you have access to garden space, grow your own**. The average American meal travels 1500 miles to get from the farm to your plate. You don't have to be a math wizard to figure out that your eating habits can contribute to an enormous amount of carbon dioxide emissions. Moreover, by buying local you

will be making a positive contribution to your community. Please check out the farmer's markets in your area.

- **Eliminate fast food from your diet.** The fast food industry creates an enormous amount of waste and does not support local producers. Besides, fast food tends to be bad for your health.
- **Become a vegetarian or a vegan**. Animal agriculture creates soil, water and air pollution and is a major contributor to global warming. Besides, the yogic practice of nonharming applies to all beings and thus to the environment as a whole.
- **Compost.** If you can't do it outdoors, try indoor vermicomposting. Check out your local library for books that can teach you how to compost properly or look for articles online.
- **Say NO to every plastic bag that is offered to you.** Make the switch over to cloth shopping bags for *all* your shopping needs, not just the grocery store. Many communities are moving toward an environment free of plastic shopping bags. Please encourage yours to do the same.
- **Plant trees**. Growing trees are carbon dioxide vacuum cleaners! Check out Trees for the Future at www.plant-trees.org. Even if you can only plant a couple of trees in your yard to offer some shade to your house, this could reduce your air conditioning needs during the summer months.
- **Make the switch to green power**. Many energy service providers have this option available, and it may only cost you a few extra dollars a month. Ask you local provider for more information.
- **Change your transportation habits.** Walk, bike, carpool, and use public transportation like the bus and train whenever possible.
- **Fly less often, if at all.** If you absolutely must fly, please purchase high-quality carbon offsets. Check the online guide

at *www.cleanair-coolplanet.org/consumerguidetocarbonoff sets.pdf.*

- **Be a real eco-traveler**. Eco-tourism is a very fast growing industry and is not necessarily environmentally friendly. Consider exploring and volunteering closer to your own home, and please remember that air-travel is still a major contributor to global warming even if your adventure has "eco" attached to it.
- **Do not use plastic water bottles**. Drink tap water that is safe to drink or, if necessary, filter your tap water. If you need to carry water with you, choose a stainless steel or glass water bottle. Commercial bottled water is not only grossly overpriced but also when packaged in plastic containers is a major contributor to land, air, and water pollution and may be hazardous to your health (owing to plastic leaking toxic phthalates).
- **Switch your investments over to socially and environmentally responsible investments and be sure that all future investments are done in the same manner**. Be proactive in learning how your money is being invested and make sure that the companies your support also address issues relating to positive changes regarding climate change.
- **Replace your incandescent light bulbs with compact fluorescent light bulbs**. They not only last about 10 times longer but also use two-thirds less energy.
- **Use only environmentally friendly cleaning products**.
- **Remove yourself from junk mail lists**. This, again, won't cost you a penny but saves c. 100 million trees, which are used every year to stuff our mailboxes with junk!
- **Encourage publishers of books, magazines, newsletters, and newspapers to use ancient forest friendly 100% postconsumer recycled paper for all their printing needs**

and congratulate those that do.

- **Practice eco-friendly flushing: If it's yellow, let it mellow. If it's brown, flush it down.** The average four-person family flushes the toilet about 20,000 times a year. Even if you are using a water-efficient toilet that consumes 1.6 gallons (6 liters) per flush, this still amounts to an incredible amount of water wasted every year.
- **Choose to have an environmentally friendly yard and garden**. Select indigenous plants and vegetation to assist in reducing the need for fertilizer, and don't use dangerous chemical pesticides. Consider purchasing a push reel mower instead of the conventional gas or electric mowers. Your neighbors and the surrounding air will be grateful to you, as will be your muscles.
- **If you are purchasing or renting a vehicle look for the most fuel-efficient and low-polluting one available and consider a plug-in hybrid.**
- **Go carbon neutral.** Going carbon neutral is a relatively easy way of taking responsibility for the negative environmental impact that we create in our daily lives. Check out *The Consumer's Guide to Retail Carbon Offset Providers* at www.cleanair-coolplanet.org/consumersguidetocarbonoffsets.pdf, and please don't use this as a substitute for simplifying the way you live.
- **Take action by writing to your political leaders and voting.** Political leaders are influenced by the public, and you can make a huge difference by simply speaking up.
- **Support your public library**. Not many things are free these days but libraries still are, and they are a great resource for books, DVDs, CDs, magazines, and newspapers, which you might otherwise have to purchase. If you don't see an item on the shelves, ask the staff if they can order it.

- **Encourage your business, organization, school, family, and friends to reduce their carbon emissions and share this list (and, perhaps, our book) with everyone you know!**

Select Organizations

David Suzuki Foundation. A science-based Canadian environmental organization with a website that is second to none for accurate, well-researched information: www.davidsuzuki.org

Environmental Health News. Provides links to articles on environmental health issues: www.environmentalhealthnews.org

Northwest Coalition for Alternatives to Pesticides. Offers news updates, action alerts, programs and various publications on alternatives to pesticides: www.pesticide.org

World Resources Institute. WRI's publications, podcasts, and articles offer in-depth treatments of issues covered in *Green Yoga*: www.wri.org

Trees for the Future. A nonprofit that initiates and supports agroforestry self-help projects in cooperation with groups and individuals in developing countries: www.plant-trees.org

Greenpeace Canada / Greenpeace USA. A global organization that campaigns for various issues—from global warming to genetic engineering: www.greenpeace.org or www.greenpeaceusa.org

World Wildlife Fund. An international organization operating in more than 100 countries, which works for a future in which humans live in harmony with nature: www.panda.org

Co-op America. Promotes environmental sustainability and social and economic justice through consumer education: www.coopamerica.org

The Union of Concerned Scientists. A group of scientists who combine independent scientific research and citizen action to help implement solutions to environmental and global security issues: www.ucsusa.org

E/The Environmental Magazine. Provides information, news, and resources for people concerned about the environment: www.emagazine.com

The Green Guide. Provides information about various environmental issues and practical advice: www. thegreenguide.com

Adbusters. An anti-consumerist organization based in Canada: www.adbusters.org

Friends of the Earth Canada or USA. Serves as a voice for the environment and for the renewal of communities and the Earth through research, education, and advocacy: www.foecanada.org or www.foe.org

Earth Island Institute. Provides organizational support in developing projects for the conservation, preservation, and restoration of the global environment: www.earthisland.org

Freecycle TM. A network providing an electronic forum for individuals and nonprofits to "recycle" unwanted items: www.freecycle.org

Lobsa. An animal rights group that offers an informative, well-researched newsletter: www.lobsa.org

Care2 Make a Difference. The largest online community for people who want to make a difference, offering news, green living section, newsletters and petitions: www.care2.com

WorldChanging. An ongoing weblog discussing and analyzing tools, ideas, models, and technologies for building a better future: www.worldchanging.com

Climate Wire. This is a leading multisectoral, international news service specifically focusing on the issue of climate change: www.climatewire.org

Seeds of Diversity. A Canadian charitable organization dedicated to the conservation, documentation and use of public-domain nonhybrid plants: www.seeds.ca

National Resources Defense Council. This is one of the most effective environmental action organizations: www.nrdc.org

One Planet, One Life. Works toward educating the public about the global environmental crisis: http://oneplanetonelife.com

The Simple Living Network. This website is filled with resources, tools, examples, and contacts for conscious, simple, healthy and restorative living: www.simpleliving.net/main/

Documentaries

Altered Oceans. A five-part series on the crisis in the seas by Kenneth R. Weiss and Usha Lee McFarling: www.latimes.com/news/local/oceans/la-oceans-series,0,7842752.special

Frontline. Must-see online documentaries: Hot Politics, Harvest of Fear (GMfoods) Kim's Nuclear Gamble, Merchants of Cool, News War, The Persuaders, and What's Up with the Weather: www.pbs.org/wgbh/pages/frontline/

Bull Frog Films. One of the best sites to find documentaries on the environment, ecology, ethics, consumerism, and animal and human rights: www.bullfrogfilms.com

Online Documents

Consumer's Guide to Retail Carbon Offset Providers — A well-researched guide that everyone should read before finding a carbon offset provider: www.cleanair-coolplanet.org/consumerguidetocarbonoffset.pdf

How to Save the Climate by Greenpeace International — An easy-to-understand document about climate change, which will appeal to all age groups: www.greenpeace.org.uk/files/pdfs/climate/howtosavetheclimatepers.pdf

Driven to Action: A Citizen's Toolkit. Produced by the David Suzuki Foundation.

Part 3: Shaping Decisions — This document describes how to be an effective lobbyist: www.davidsuzuki.org/files/climate/ontario/shapingdecisions.pdf

Part 4: Working With the Media — This document describes how to get media attention for environmental projects: www.davidsuzuki.org/files/climate/ontario/sprawl-media-tips.pdf

Recommended Reading

Choose to Reuse: An Encyclopedia of Services, Products, Programs and Charitable Organizations that Foster Reuse by Nikki & David Goldbeck. ▪ *Beyond Recycling: A Re-user's Guide: 336 Practical Tips to Save Money and Protect the Environment* by Kathy Stein. ▪ *The 100-Mile Diet: A Year of Local Eating* by Alisa Smith and J. B. Mackinnon. ▪ *Low Carbon Diet: A 30 Day Program to Lose 5000 Pounds* by David Gershon. ▪ *Ecoholic: Your Guide to the Most Environmentally Friendly Information, Products and Services in Canada* by Adria Vasil. ▪ *Clean and Green* by Annie Berthold-Bond. ▪ *EarthScore: Your Personal Environmental Audit and Guide* by Donald W. Lotter. ▪ *The Consumer's Guide to Effective Environmental Choices: Practical Advice from the Union of Concerned Scientists* by Michael Brower and Warren Leon. ▪ *The Better World Handbook: From Good Inventions to Everyday Actions* by Ellis Jones et al. ▪ *Homes That Heal (and Those That Don't): How Your Home Could be Harming Your Family's Health* by Athena Thompson. ▪ *Rodale's All-New Encyclopedia of Organic Gardening: The Indispensable Resource for Every Gardener* ed. by Fern Marchall Bradley and Barbara W. Ellis. ▪ *Green Living* by E/ The Environmental Magazine.

EPILOGUE

This book is necessarily filled with facts and figures. Our basic proposition that we are in the throes of a worldwide crisis should by now have become self-evident to every single reader. We realize that some of our readers—those who did not read our book carefully or ponder its evidence in an unbiased way—might deem our message too radical.

We think, however, that our message is commensurate with the weight of the evidence. Many of those working at the forefront of science, especially biologists, climatologists, glaciologists, and hydrologists, are actually scared by the implications of their own findings. Based on the available scientific evidence, we (and many others) have come to believe that humanity has *no* time left for idle motion or for business as usual. In the words of the American cultural historian Thomas Berry:

> The day of reckoning has come. In this disintegrating phase of our industrial society, we now see ourselves not as the splendor of creation, but as the most pernicious mode of earthly being. We are the termination, not the fulfillment of the earth process. If there were a parliament of creatures, its first decision might well be to vote the humans out of the community, too deadly a presence to tolerate any further.*

*Thomas Berry, *The Dream of the Earth* (San Francisco: Sierra Club Books, 1985), p. 209.

Collectively and individually, we must *now* make radical lifestyle changes. We must simplify *greatly* and adopt environment-friendly ways of living. Humanity's future and indeed the future of life on Earth is at stake. It would also be wise to prepare for the inevitable repercussions of global warming. These will be severe and will predictably cause tremendous suffering throughout the world. No one will be exempt from the drastic effects of Nature's rebalancing mechanisms.

Above all, we must cultivate great compassion for our fellow beings, including all nonhuman creatures with whom we share this precious planet and upon whom we have inflicted so much suffering because of our profligate and largely irresponsible lifestyle.

As humanity's spiritual prodigies have amply demonstrated throughout known history, as human beings we are in principle capable of great self-transformation. Usually during times of war or communal crisis, we manifest some of our intrinsic potential for creativity, altruism, solidarity, and self-transcending action. Global warming is a collective crisis of unprecedented magnitude. When this planetary challenge has been properly understood as a threat to our survival, we may yet find within us the wisdom and strength to drop our petty self-centered concerns in favor of decisive action that will benefit our planet at large.

We are calling upon all Yoga practitioners to intensify their practice by becoming *viras,* or "heroes," in the old sense of the word, who put the common weal before their own consumerist comfort and uninspected predilections. The time has come to *live* Yoga with as much heartiness and genuineness as we can possibly muster. If Yoga practitioners won't respond to this unique and perilous crisis, who will?

Let us use this planetary crisis as a unique opportunity to grow as individuals and as a species.

SELECT BIBLIOGRAPHY

Especially important publications are marked with an asterisk (*). Further readings are given in the footnotes.

*__Brown__, Lester R. *Plan B 2.0.* New York: W. W. Norton, 2006.

__________. *Eco-Economy: Building an Economy for the Earth.* New York: W. W. Norton, 2001.

*__Dauncey__, Guy with Patrick Mazza. *Stormy Weather: 101 Solutions to Global Climate Change.* Gabriola Island, B.C.: New Society Publishers, 2001.

*__Dobson__, Charles. *The Troublemaker's Teaparty: A Manual for Effective Citizen Action.* Gabriola Island, B.C.: New Society Publishers, 2003.

Drengson, Alan. *Deep Ecology Movement: An Introductory Anthology.* Berkeley, Calif.: North Atlantic Books, 1995.

*__Elgin,__ Duane. *Voluntary Simplicity: Toward a Way of Life That is Outwardly Simple, Inwardly Rich.* New York: Morrow, 1993.

*__Feuerstein__, Georg. *Yoga Morality: Ancient Teachings At a Time of Global Crisis.* Prescott, Az.: Hohm Press, 2007.

__________. *The Yoga Tradition: Its History, Philosophy, Literature, and Practice.* Prescott, Ariz.: Hohm Press, 2d ed. 2001.

__________. *Aha! Reflections on the Meaning of Everything.* Eastend, Sask.: Traditional Yoga Studies, 2007.

*__Flannery__, Tim. *The Weather Makers: How We Are Changing the Climate And What It Means For Life On Earth.* New York: HarperCollins, 2005.

Fujita, Rod. *Heal the Ocean: Solutions For Saving Our Seas.* Gabriola Island, B.C.: New Society Publishers, 2003.

Gelbspan, Ross. *The Heat Is On: The High Stakes Battle Over Earth's Threatened Climate.* Reading, Mass.: Addison-Wesley, 1997.

Halweil, Brian and Lisa Mastny, eds. *State of the World 2004: Special Focus: The Consumer Society.* New York: W. W. Norton, 2004.

Heinberg, Richard. *The Party's Over: Oil, War and the Fate of Industrial Societies.* Gabriola Island, B.C.: New Society Publishers, 2003.

__________. *Cloning the Buddha: The Moral Impact of Biotechnology.* Wheaton, Ill.: Quest Books, 1999.

Jamieson, Dale, ed. *A Companion to Environmental Philosophy.* Malden, Mass.: Blackwell, 2003.

Jensen, Derrick. *A Language Older Than Words.* White River Junction, Vt.: Chelsea Green, 2004.

Kaza, Stephanie and Kenneth Kraft, eds., *Dharma Rain: Sources of Buddhist Environmentalism.* Boston, Mass.: Shambhala Publications, 2000.

Meadows, Donella, Jorgan Randers, and Dennis Meadows. *Limits to Growth: The 30-Year Update.* White River Junction, Vt.: Chelsea Green, 2004.

Merkel, Jim. *Radical Simplicity: Small Footprints on a Finite Earth.* Gabriola Island, B.C.: New Society Publishers, 2003.

*__Naess,__ Arne. *Ecology, Community and Lifestyle: Outline of an Ecosophy.* Cambridge: Cambridge University Press, repr. 1990.

__________. *The Selected Works of Arne Naess.* Adapted by Alan Drengson. New York: Springer, 2005. 10 vols.

Nattrass, Brian and Mary Altomare. *Dancing With the Tiger: Learning Sustainability Step By Natural Step.* Gabriola Island, B.C.: New Society Publishers, 2003.

Renner, Michael et al., eds. *State of the World 2005: Redefining Global Security.* New York: W. W. Norton, 2005.

*__Robbins__, John. *A New Diet for America.* Tiberon, Calif.: H. J. Kramer, repr. 1998.

Sessions, George ed., *Deep Ecology For the 21st Century.* Boston and London: Shambhala Publications, 1995.

*__Shiva__, Vandana. *Biopiracy: The Plunder of Nature and Knowledge.* Cambridge, Mass.: South End Press, 1997.

Sloman, Lynn. *Car Sick: Solutions For Our Car-Addicted Culture.* White River Junction, Vermont: Chelsea Green, 2006.

*__Suzuki__, David with Amanda McConnell. *The Sacred Balance.* Vancouver, British Columbia: Greystone Books, 2002.

*__________ and Holly Dressel, *Good News For a Change: How Everyday People Are Helping the Planet.* Vancouver: Greystone Books, 2003.

Thompson, Athena. *Homes That Heal And Those That Don't.* Gabriola Island, B.C.: New Society Publishers, 2004.

INDEX

ABOUT THE AUTHORS

Georg Feuerstein, Ph.D., has authored more than thirty books, including the award-winning *Shambhala Encyclopedia of Yoga, The Yoga Tradition, Yoga Morality,* and *Aha! Reflections on the Meaning of Everything,* and *Transparent Leaves From the Tree of Life: Metaphysical Poems.* His distance-learning courses on the philosophy, history, and literature of Yoga are available through TYS (www.traditionalyogastudies.com). In 2004, after relocating from the U.S. to Canada, he decided to shelve his indological work and focus instead on urgent environmental and social issues.

Brenda Feuerstein is a former music teacher and Yoga instructor, as well as health and fitness consultant. She has for many years been active in the fields of tree planting, alternative energy and housing. Today she serves as director of TYS and is codirector of Green Yoga Initiative and coeditor of GYI's e-newsletter.

Georg and Brenda live in Saskatchewan, where they enjoy organic gardening and, whenever their time allows, hiking in the Prairies. They only travel by car or plane when absolutely necessary and otherwise live a simple life dedicated to educating the public about the imminent danger of environmental collapse through their writings and website.

Together they are currently preparing *Green Dharma* and *Green Yoga Workbook.*